Offprint from
Numerische Mathematik, 14, 79—218 (1969)

Springer-Verlag Berlin Heidelberg GmbH

Report on the Algorithmic Language ALGOL 68

A. van Wijngaarden (Editor)
B. J. Mailloux, J. E. L. Peck and C. H. A. Koster

ISBN 978-3-662-38646-0 ISBN 978-3-662-39502-8 (eBook)
DOI 10.1007/978-3-662-39502-8

Report of IFIP TC 2 to the General Assembly of IFIP in the matter of the Algorithmic Language ALGOL 68

Attached will be found the first Report on the Algorithmic Language ALGOL 68, submitted with a recommendation for publication under IFIP auspices by Technical Committee 2 in response to a request from Working Group 2.1 (ALGOL). This Working Group has been directed, by TC 2, for a number of years towards concern for the design of common programming languages. During this time, TC 2 has realized the magnitude and difficulty of this task.

In the course of pursuing its responsibilities, WG 2.1 commissioned and has guided the work of the four listed authors of the attached Report and has stated its desire to acknowledge the considerable effort devoted by these men to their task. This Report, however, must be regarded as constituting more than simply the work of four authors. Its content has been influenced throughout and the results are, in the main, a consequence of discussions within the Working Group.

The Report is, thus, submitted by TC 2 as representing the current consolidated outcome of WG 2.1 activity. This does not imply that any member of the Working Group necessarily agrees with every aspect of the undertaking nor guarantees that all relevant aspects of the problem have been considerated. Indeed, fear has been expressed, in the form of a Minority Report from the Working Group to TC 2, that the direction taken in the attached Report will not lead towards the goal of providing appropriate programming tools for the future. It has, however, been decided by the Working Group, that this work has reached the proper stage for submission to the crucial tests of implementation and subsequent use by the computing community. In this opinion, TC 2 concurs. Therefore, this Report is submitted for publication as representing one of the possible approaches to the subject, rather than in the spirit of a final answer.

WG 2.1 will be directed to keep continually under review experience obtained as a consequence of this publication, so that it may institute such corrections and revisions to the Report as become desirable. To this end, TC 2 adds its endorsement to the request of WG 2.1 that all who wish to contribute to this work should do so through the established medium of the ALGOL-Bulletin.

In consonance with these declared aims, TC 2 requests that the IFIP General Assembly take those actions necessary and proper to insure expeditious and widespread publication of this Report on the Algorithmic Language ALGOL 68 together with this page as a covering statement.

Numer. Math. 14, 79–218 (1969)

Report on the Algorithmic Language ALGOL 68 *

A. VAN WIJNGAARDEN (Editor), B. J. MAILLOUX, J. E. L. PECK and C. H. A. KOSTER

Acknowledgements

{*Habent sua fata libelli.*
De litteris, *Terentianus Maurus.*}

Working Group 2.1 on ALGOL of the International Federation for Information Processing has discussed the development of "ALGOL X", a successor to ALGOL 60 [3] since 1963. At its meeting in Princeton in May 1965, WG 2.1 invited written descriptions of the language based on the previous discussions. At the meeting in St Pierre de Chartreuse near Grenoble in October 1965, three reports describing more or less complete languages were amongst the contributions, by Niklaus Wirth [8], Gerhard Seegmüller [6], and Aad van Wijngaarden [9]. In [6] and [8], the descriptional technique of [3] was used, whereas [9] featured a new technique for language design and definition. Other significant contributions available were papers by Tony Hoare [2] and Peter Naur [4, 5].

At meetings in Kootwijk near Amsterdam in April 1966, Warsaw in October 1966, Zandvoort near Amsterdam in May 1967, Tirrenia near Pisa in June 1968, North Berwick near Edinburgh in July 1968 and Munich in December 1968, a number of successive approximations to a final report were submitted by a team working in Amsterdam, consisting first of A. van Wijngaarden and Barry Mailloux [10], later reinforced by John Peck [11, 12], and finally by Kees Koster [13, 14, 15, 16]. Versions were used during courses on the language held in Amsterdam [12], Bakuriani [13], Grenoble [13], Copenhagen [14], Esztergom [14], Calgary [14] and Chicago [16]. These courses served as test cases and the experience gained in explaining the language to skilled audiences and the reactions of the students influenced the succeeding versions.

The authors acknowledge with pleasure and thanks the whole-hearted cooperation, support, interest, criticism and violent objections from members of WG 2.1 and many other people interested in ALGOL. At the risk of embarrassing omissions, special mention should be made of Jan Garwick, Jack Merner, Peter Ingerman and Manfred Paul for [1], the Brussels group consisting of M. Sintzoff, P. Branquart, J. Lewi and P. Wodon for numerous brainstorms, A. J. M. van Gils of Apeldoorn, G. Goos and his group of Munich, also for [7], G. S. Tseytin of Leningrad and L. G. L. T. Meertens and J. W. de Bakker of Amsterdam. An occasional choice of a, not inherently meaningful, identifier in the sequel may compensate for not mentioning more names in this section.

1. Garwick, J. V., Merner, J. M., Ingerman, P. Z., Paul, M.: Report of the ALGOL-X-I-O subcommittee, WG 2.1 working paper, July 1966.
2. Hoare, C. A. R.: Record handling, WG 2.1 working paper, October 1965; also AB.21.3.6, November 1965.

* This Report has been reviewed by Technical Committee 2 on Programming Languages and approved for publication by the General Assembly of the International Federation for Information Processing. Reproduction of the Report, for any purpose, but only of the whole text together with the cover-letter, is explicitly permitted without formality.

3. Naur, P. (Editor): Revised report on the algorithmic language ALGOL 60, Regnecentralen, Copenhagen, 1962, and elsewhere.
4. — Proposals for a new language, AB.18.3.9, October 1964.
5. — Proposal for introduction on aims. WG 2.1 working paper, October 1965.
6. Seegmüller, G.: A proposal for a basis for a report on a successor to ALGOL 60, Bavarian Acad. Sci., Munich, October 1965.
7. Goos, G., Scheidig, H., Seegmüller, G., Walther, H.: Another proposal for ALGOL 67, Bavarian Acad. Sci., Munich, May 1967.
8. Wirth, N.: A proposal for a report on a successor of ALGOL 60, Mathematisch Centrum, Amsterdam, MR 75, October 1965.
9. Wijngaarden, A. van: Orthogonal design and description of a formal language, Mathematisch Centrum, Amsterdam, MR 76, October 1965.
10. — Mailloux, B. J.: A draft proposal for the algorithmic language ALGOL X, WG 2.1 working paper, October 1966.
11. — — Peck, J. E. L.: A draft proposal for the algorithmic language ALGOL 67, Mathematisch Centrum, Amsterdam, MR 88, May 1967.
12. — — — A draft proposal for the algorithmic language ALGOL 68, Mathematisch Centrum, Amsterdam, MR 92, November 1967.
13. — (Editor), Mailloux, B. J., Peck, J. E. L., Koster, C. H. A.: Draft report on the algorithmic language ALGOL 68, Mathematisch Centrum, MR 93, January 1968.
14. — — — — Working document on the algorithmic language ALGOL 68, Mathematisch Centrum, Amsterdam, MR 95, July 1968.
15. — — — — Penultimate draft report on the algorithmic language ALGOL 68, Mathematisch Centrum, MR 99, October 1968.
16. — (Editor), Mailloux, B. J., Peck, J. E. L., Koster, C. H. A.: Final draft report on the algorithmic language ALGOL 68, Mathematisch Centrum, Amsterdam, MR 100, December 1968.

Contents

0. Introduction

0.1. Aims and Principles of Design

a) In designing the Algorithmic Language ALGOL 68, Working Group 2.1 on ALGOL of the International Federation for Information Processing expresses its belief in the value of a common programming language serving many people in many countries.

b) ALGOL 68 is designed to communicate algorithms, to execute them efficiently on a variety of different computers, and to aid in teaching them to students.

c) The members of the Group, influenced by several years of experience with ALGOL 60 and other programming languages, accepted the following as their aims:

0.1.1. Completeness and Clarity of Description

The Group wishes to contribute to the solution of the problems of describing a language clearly and completely. The method adopted in this Report is based upon a strict language comprizing a language core, whose description is minimized. The remainder of the language is described in terms of this core, thereby reducing the amount of semantic description at the cost of a heavier burden on the syntax. It is recognized, however, that this method may be difficult for the uninitiated reader. Therefore, a companion volume, entitled "Informal Introduction to ALGOL 68", has been prepared at the request of the Group by C. H. Lindsey and S. G. van der Meulen, and further companion volumes on specific aspects of the language may well follow.

0.1.2. Orthogonal Design

The number of independent primitive concepts was minimized in order that the language be easy to describe, to learn, and to implement. On the other hand, these concepts have been applied "orthogonally" in order to maximize the expressive power of the language, and yet without introducing deleterious superfluities.

0.1.3. Security

ALGOL 68 has been designed in such a way that nearly all syntactical and many other errors can be detected easily before they lead to calamitous results. Furthermore, the opportunities for making such errors are greatly restricted.

0.1.4. Efficiency

ALGOL 68 allows the programmer to specify programs which can be run efficiently on present-day computers and yet do not require sophisticated and time-consuming optimization features of a compiler; see, e.g., 11.8.

0.1.4.1. Static Mode Checking

The syntax of ALGOL 68 is such that no mode checking during run time is necessary except during the elaboration of conformity relations, the use of which is required only in those cases in which the programmer explicitly makes use of the flexibility offered by the united mode feature.

0.1.4.2. Static Scope Checking

The syntax of ALGOL 68 is such that no scope checking during run time is necessary except in some cases in which the programmer explicitly makes use of the flexibility offered by the absence of syntactical scope restrictions.

0.1.4.3. Mode-Independent Parsing

The syntax of ALGOL 68 is such that the parsing of a program can be performed independently of the modes of its constituents. Moreover, there is an algorithm which determines in a finite number of steps whether an arbitrary given sequence of symbols is a proper program.

0.1.4.4. Independent Compilation

The syntax of ALGOL 68 is such that the main line programs and procedures can be compiled independently of one another without loss of object-program efficiency, provided that during each such independent compilation, specification of the mode of all nonlocal quantities is provided; see the remarks after 2.3.c.

0.1.4.5. Loop Optimization

Iterative processes are formulated in ALGOL 68 in such a way that straightforward application of well-known optimization techniques yields large gains during run time without excessive increase of compilation time.

0.1.4.6. Representations

Representations of ALGOL 68 symbols have been chosen so that the language may be implemented on computers with a minimal character set. At the same time implementers may take advantage of a larger character set, if it is available.

0.2. Comparison with ALGOL 60

a) ALGOL 68 is a language of wider applicability and power than ALGOL 60. Although influenced by the lessons learned from ALGOL 60, ALGOL 68 has not been designed as an expansion of ALGOL 60 but rather as a completely new language based on new insights into the essential, fundamental concepts of computing and a new description technique.

b) The result is that the successful features of ALGOL 60 reappear in ALGOL 68 but as special cases of more general constructions, along with completely new features. It is, therefore, difficult to isolate differences between the two languages; however, the following sections are intended to give insight into some of the more striking differences.

0.2.1. Values in ALGOL 68

a) Whereas ALGOL 60 has values of the types **integer**, **real** and **Boolean**, ALGOL 68 features an infinity of "modes", i.e., generalizations of the concept "type".

b) Each plain value is either arithmetic, i.e., of integral or real mode and then it is of one of several lengths, or it is of boolean or character mode.

c) In ALGOL 60, composition of values is possible into arrays, whereas in ALGOL 68, in addition to such "multiple" values, also "structured" values, composed of values of possibly different modes, are defined and manipulated. An example of a multiple value is the character array, which corresponds approximately to the ALGOL 60 string; examples of structured values are complex numbers, machine words considered as sequences of bits or of bytes, and symbolic formulae.

d) In ALGOL 68, the concept of a "name" is introduced, i.e., a value which is said to "refer to" another value; such a name-value pair corresponds to the ALGOL 60 variable. However, any name may take the value position in a name-value pair and thus chains of indirect addresses can be built up.

e) The ALGOL 60 concept of procedure body is generalized in ALGOL 68 to the concept of "routine", which includes also the formal parameters, and which is itself a value and therefore can be manipulated like any other value; the ALGOL 68 concept "format" has no ALGOL 60 counterpart.

f) In contrast with plain values and multiple and structured values composed of plain values only, the significance of a name, routine or format or of a multiple or structured value composed of names, routines or formats, possibly amongst other values, is, in general, dependent on the context in which it appears. Therefore, the use of names, routines and formats is subject to some natural restrictions related to their "scope".

0.2.2. Declarations in ALGOL 68

a) Whereas ALGOL 60 has type declarations, array declarations, switch declarations and procedure declarations, ALGOL 68 features the "identity declaration" whose expressive power includes all of these, and more. In fact, the identity declaration declares not only variables, but also constants, of any mode, and, moreover, forms the basis of a highly efficient and powerful parameter mechanism.

b) Moreover, in ALGOL 68, a "mode declaration" permits the construction of new modes from already existing ones. In particular, the modes of multiple values and of structured values may be defined this way; in addition, a union of modes may be defined for use in an identity declaration allowing each value referred to by a given name to be of any one of the uniting modes.

c) Finally, in ALGOL 68, a "priority declaration" and an "operation declaration" permit the introduction of new operators, the definition of their operation and the extension or revision of the class of operands applicable to already established operators.

0.2.3. Dynamic Storage Allocation in ALGOL 68

Whereas ALGOL 60 (apart from the so-called "own dynamic arrays") implies a "stack"-oriented storage-allocation regime, sufficient to cope with a statically (i.e., at compile time) determined number of single or multiple values, ALGOL 68 provides, in addition, the ability to generate a dynamically (i.e., at run time) determined number of values, which ability implies the use of additional storage-allocation techniques.

0.2.4. Collateral Elaboration in ALGOL 68

Whereas, in ALGOL 60, statements are "executed consecutively", in ALGOL 68, "phrases" are "elaborated serially" or "collaterally". This last facility is conducive to more efficient object programs under many circumstances, and increases the expressive power of the language. Facilities for parallel programming, though restricted to the essentials in view of the none-too-advanced state of the art, have been introduced.

0.2.5. Standard Declarations in ALGOL 68

The ALGOL 60 standard functions are all included in ALGOL 68 along with many other standard declarations. Amongst these are "environment enquiries", which make it possible to determine certain properties of an implementation, and "transput" declarations, which make it possible, at run time, to obtain data from and to deliver results to external media.

0.2.6. Some Particular Constructions in ALGOL 68

a) The ALGOL 60 concepts of block, compound statement and parenthesized expressions are unified in ALGOL 68 into "closed clause". A closed clause may be an expression and possess a value. Similarly, the ALGOL 68 "assignation", which is a generalization of the ALGOL 60 assignment statement, may be an expression and, as such, also possess a value.

b) The ALGOL 60 concept of subscription is generalized to the ALGOL 68 concept of "indexing", which allows the selection not only of a single element of an array but also of subarrays with the same or any smaller dimensionality and with possibly altered bounds.

c) ALGOL 68 provides not only the multiple values mentioned in 0.2.1.c, but also "collateral expressions" which serve to compose these values or structured values from other, simpler values.

d) The ALGOL 60 for statement is modified into a more concise and efficient "repetitive statement".

e) The ALGOL 60 conditional expression and conditional statement, unified into a "conditional clause", are improved by requiring them to end with a closing symbol whereby the two alternative clauses admit the same syntactic possibilities. Moreover, the conditional clause is generalized into a "case clause" which allows the efficient selection from an arbitrary number of clauses depending on the value of an integral expression or of a conformity relation.

f) Some less successful ALGOL 60 concepts, such as own quantities and integer labels, have not been included in ALGOL 68, and some concepts like designational expressions and switches do not appear as such in ALGOL 68, but their expressive power is included in other, more general, constructions.

{True wisdom knows it must comprise some nonsense as a compromise, lest fools should fail to find it wise.
Grooks, *Piet Hein}.*

1. Language and Metalanguage

1.1. The Method of Description

1.1.1. The Strict, Extended and Representation Languages

a) ALGOL 68 is a language in which "programs" can be formulated for "computers", i.e., "automata" or "human beings". It is defined in three stages, the "strict language", the "extended language" and the "representation language".

b) In the definition, the "English language" and a "formal language" are used. For both of these languages, as also for the strict language and the extended language, typographical and syntactic marks are used which bear no immediate relation to those used for the representation language.

1.1.2. The Syntax of the Strict Language

a) The strict language is defined by means of a "syntax" and "semantics". This syntax is a set of "production rules" for "notions"; it is expressed in

"small syntactic marks", in this Report "a", "b", "c", "d", "e", "f", "g", "h", "i", "j", "k", "l", "m", "n", "o", "p", "q", "r", "s", "t", "u", "v", "w", "x", "y" and "z";

"large syntactic marks", in this Report "A", "B", "C", "D", "E", "F", "G", "H", "I", "J", "K", "L", "M", "N", "O", "P", "Q", "R", "S", "T", "U", "V", "W", "X", "Y" and "Z";

"other syntactic marks", in this Report "." ("point"), "," ("comma"), ":" ("colon"), ";" ("semicolon") and "*" ("asterisk").

{Note that these marks are in another type font than the marks in this sentence.}

b) A "protonotion" is a nonempty, possibly infinite, sequence of small syntactic marks; a notion is a protonotion for which there is a production rule; a "symbol" is a protonotion ending with 'symbol'.

c) A production rule for a given notion consists of that notion, possibly preceded by an asterisk, followed by a colon, a "list of notions" {see d}, and a point, in that order. The list of notions is said to be a "direct production" of the given notion.

d) A list of notions is a nonempty sequence of "members" separated by commas; a member is either a notion and is then said to be "productive" {, or nonterminal,} or is a symbol {, which is terminal,} or is empty, or is some other protonotion {and then the production rule of whose list of notions it is a member is said to be a "blind alley"}.

e) A "production" of a given notion is either a direct production of that given notion or a list of notions obtained by replacing a productive member in some production of the given notion by a direct production of that productive member.

f) A "terminal production" of a notion is a production of that notion each of whose members is either a symbol or empty.

{In the production rule

'variable point numeral : integral part option, fractional part.'

(5.1.2.1.b) of the strict language, the list of notions

'integral part option, fractional part'

is a direct production of the notion 'variable point numeral', and contains two members, both of which are productive. A terminal production of this same notion is

'digit zero symbol, point symbol, digit one symbol'.

The member 'digit zero symbol' is an example of a symbol, and is terminal. The protonotion 'twas brillig and the slithy toves' is neither a symbol nor a notion in the sense of this Report, in that it does not end with 'symbol' and no production rule for it is given (1.1.5.b,c).}

1.1.3. The Syntax of the Metalanguage

a) Some production rules of the strict language are given explicitly and others are obtained with the aid of a "metalanguage" whose syntax is a set of production rules for "metanotions".

b) A metanotion is a nonempty sequence of large syntactic marks.

c) A production rule for a given metanotion consists of that metanotion followed by a colon, a "list of metanotions" {see d}, and a point, in that order. The list of metanotions is said to be a direct production of the given metanotion.

d) A list of metanotions is a nonempty sequence of "metamembers" separated by blanks; a metamember is either a metanotion and is then said to be productive, or a, possibly empty, sequence of small syntactic marks.

e) A production of a given metanotion is either a direct production of that given metanotion or a list of metanotions obtained by replacing a productive metamember in some production of the given metanotion by a direct production of that productive metamember.

f) A terminal production of a metanotion is a production of that metanotion none of whose metamembers is productive.

{In the production rule 'TAG : LETTER.', derived from 1.2.1.r, 'LETTER' is a direct production of the metanotion 'TAG', and consists of one metamember which is productive. A particular terminal production of the metanotion 'TAG' is 'letter x' (see 1.2.1.s,t). In the production rule 'EMPTY : .' (1.2.1.i), the metanotion 'EMPTY' has a direct production which consists of one empty metamember.}

1.1.4. The Production Rules of the Metalanguage

The production rules of the metalanguage are the rules obtained from the rules in Section 1.2 in the following steps:

Step 1: If some rule contains one or more semicolons, then it is replaced by two new rules, the first of which consists of the part of that rule up to and including the first semicolon with that semicolon replaced by a point, and the second of which consists of a copy of that part of the rule up to and including the colon, followed by the part of the original rule following its first semicolon, whereupon Step 1 is taken again;

Step 2: A number of production rules for the metanotion 'ALPHA' {1.2.1.t}, each of whose direct productions is another small syntactic mark, may be added.

{For instance, the rule

'TAG : LETTER ; TAG LETTER ; TAG DIGIT.',

from 1.2.1.r, is replaced by the rules

'TAG : LETTER.' and 'TAG : TAG LETTER ; TAG DIGIT.',

and the second of these is replaced by

'TAG : TAG LETTER.' and 'TAG : TAG DIGIT.',

thus resulting in three rules from the original one.

The reader might find it helpful to read ":" as "may be a", "," as "followed by a", and ";" as "or a".}

1.1.5. The Production Rules of the Strict Language

a) A production rule of the strict language is any rule obtained in the following steps from the rules given in Chapters 2 up to 8 inclusive in the sections whose heading is, or begins with, "Syntax":

Step 1: Identical with Step 1 of 1.1.4;

Step 2: One of the rules obtained is considered;

Step 3: If the considered rule contains one or more metanotions, then for some terminal production of such a metanotion, a new rule is obtained by replacing that metanotion, throughout a copy of the considered rule, by that terminal production, whereupon this new rule is considered instead and Step 3 is taken; otherwise, all blanks in the considered rule are removed and the rule so obtained is a production rule of the strict language.

b) A number of production rules may be added for 'indicant', 'dyadic indicant' and 'monadic indicant' {4.2.1.b,e,f}, each of whose direct productions is a symbol different from any symbol given in this Report, with the restriction that no terminal production of 'indicant' is also a terminal production of 'monadic indicant'.

c) A number of production rules may be added for 'other comment item' {3.0.9.c} and 'other string item' {5.1.4.1.b} each of whose direct productions is a symbol different from any terminal production of 'character token' with the restrictions that no terminal production of 'other comment item' is 'comment symbol' and no terminal production of 'other string item' is 'quote symbol'.

{The rule

'actual LOWPER bound : strict LOWPER bound, flexible symbol option.'

derived from 7.1.1.t by Step 1 and considered in Step 2 is used in Step 3 to provide either one of the following two production rules of the strict language:

'actuallowerbound:strictlowerbound,flexiblesymboloption.' and

'actualupperbound:strictupperbound,flexiblesymboloption.';

however, to ease the burden on the reader, who may more easily ignore blanks himself, some blanks will be retained in the symbols, notions and production rules in the rest of this Report. Thus, the rules will be written in the more readable form

'actual lower bound : strict lower bound, flexible symbol option.' and
'actual upper bound : strict upper bound, flexible symbol option.'.

Note that

'actual lower bound : strict upper bound, flexible symbol option.'

is not a production rule of the strict language, since the replacement of the metanotion 'LOWPER' by one of its productions must be consistent throughout.

Since some metanotions have an infinite number of terminal productions, the number of notions in the strict language is infinite and the number of production rules for a given notion may be infinite; moreover, since some metanotions have terminal productions of infinite length, some notions are infinitely long. For examples see 4.1.1 and 8.5.2.2. These infinities should not worry the reader. From the sequel it follows that in any program only a finite number of notions and production rules are involved, and that although notions of infinite length may be involved, their constitution is always determined by a simple substitution process defined by a finite number of symbols, viz., the program; it is this process rather than its hypothetical outcome that matters.

Some production rules obtained from a rule containing a metanotion may be blind alleys in the sense that no production rule is given for some member to the right of the colon even though it is not a symbol.}

1.1.6. The Semantics of the Strict Language

a) A terminal production of a notion is considered as a linearly ordered sequence of symbols. This order is termed the "textual order", and "following" ("preceding") stands for "textually immediately following" ("textually immediately preceding") in the rest of this Report. Typographical display features, such as blank space, change to a new line, and change to a new page do not influence this order {but see 3.1.2.b}.

b) A sequence of symbols (A protonotion) consisting of a second sequence of symbols (a second protonotion) preceded and/or followed by (a) nonempty sequence(s) of symbols (of small syntactic marks) "contains" that second sequence of symbols (second protonotion).

c) A "paranotion" when not in a section whose heading is, or begins with, "Syntax", not between "apostrophe"s ("'") and not contained in another paranotion "denotes" some number of protonotions, its "originals". A paranotion is either

i) a symbol and then it denotes itself {, e.g., "begin symbol" denotes "begin symbol"}, or

ii) a notion whose production rule does (rules do) not begin with an asterisk, and then it denotes itself {, e.g., "plusminus" denotes "plusminus"}, or

iii) a notion whose production rule does (rules do) begin with an asterisk, and then it denotes any of its direct productions {, which, in this Report, always is a notion or a symbol, e.g., "trimscript" (8.6.1.1.j) denotes "trimmer" or "subscript"}, or

iv) a paranotion in which one or more "hyphen"s ("-") have been inserted and then it denotes the originals of that paranotion before the inserting {, e.g., "begin-symbol" denotes what "begin symbol" denotes}, or

v) a paranotion followed by "s" or a paranotion ending with "y" in which that "y" has been replaced by "ies" and then it denotes some number of the originals of that paranotion before the modification {, e.g., "trimscripts" denotes some number of "trimmer"s and/or "subscript"s and "primaries" denotes some number of the notions denoted by "primary"}, or

vi) a paranotion whose first small syntactic mark has been replaced by the corresponding large syntactic mark, and then it denotes the originals of that paranotion before the modification {, e.g., "Identifiers" denotes the notions denoted by "identifiers"}, or

vii) a paranotion in which a terminal production of 'SORT' and/or of 'SOME' and/or of 'MOID' has been omitted, and then it denotes the originals of any other paranotion from which the given paranotion could be obtained by omitting a terminal production of 'SORT' and/or of 'SOME' and/or of 'MOID' {, e.g., "jump" denotes the notions denoted by "MOID jump" (8.2.7.1.c), "declaration" denotes the notions denoted by "SOME declaration" (6.2.1.a,7.0.1.a) and "clause" denotes the notions denoted by "SORTETY SOME MOID clause" (6.1.1.a, 6.2.1.b,c,d,f,6.3.1.a,6.4.1.a,c,d,e,8.1.1.a), where "SORTETY" ("SOME", "MOID") stands for any terminal production of the metanotion 'SORTETY' ('SOME', 'MOID')}.

{As an aid to the reader, paranotions, when not under Syntax or between apostrophes, are provided with hyphens where, otherwise, they are provided with blanks. Rules beginning with an asterisk have been included in order to shorten the semantics.}

d) Except as otherwise specified {f,g}, a paranotion stands for any "occurrence" of any symbol denoted by it and/or of any terminal production of any notion denoted by it considered as a terminal production of, specifically, that notion.

e) An occurrence *O* of a terminal production of a notion *N* is produced by a tree of specific productions; for this "production tree" are defined:

a "direct descendent" of *N*, i.e., a member of the direct production of *N*;

a "descendent" of *N*, i.e., a direct descendent of either *N* or a descendent of *N*;

the "offspring" of a descendent *D* of *N*, i.e., if *D* is a notion (symbol), then the occurrence of the terminal production of (the occurrence of) *D* {which is, or is contained in, *O*}; and

a "direct constituent" of *O*, i.e., the offspring of a direct descendent of *N*.

f) A paranotion *P* which is said to be a direct constituent of a paranotion *Q*, stands for all occurrences of symbols denoted by *P* or of terminal productions of notions denoted by *P* which are direct constituents of occurrences of terminal productions of notions stood for by *Q*.

A paranotion *P* is a descendent of a paranotion *Q* if it is a direct constituent of *Q* or if it is a direct constituent of a descendent of *Q*.

A paranotion *P* is a "constituent" of a paranotion *Q* if it is a descendent of *Q* but not a descendent of a descendent *R* of *Q* where *R* is the same protonotion as either *P* or *Q*.

{Hence, *1:2* is a constituent actual-row-of-rower of the actual-declarer *[1 : 2]* **struct** *([3 : 4]* **real** *a)*, but *3 : 4* is not, since it is a descendent of the descendent actual-declarer *[3 : 4]* **real**. Also, *j := 1* is a constituent assignation (8.3.1.1.a) of the assignation *i := j := 1*, but not of the serial-clause *i := j := 1; k := 2* nor of the assignations *j := 1* and *k := i := j := 1*. The assignation *j := 1* is not a direct constituent of the assignation *i := j := 1*, but the source *j := 1* is (8.3.1.1.a).

g) A paranotion which is a direct constituent of a second paranotion is a paranotion of that second paranotion.

{This permits the abbreviation of "direct constituent of", which would appear frequently under Semantics, to "of", "its" or even "the", e.g., in the assignation (8.3.1.1.a) *i := 1*, *i* is its destination or *i* is the, or a, destination of the assignation *i := 1*, whereas *i* is a constituent, but not simply a, destination of the serial-clause *i := 1; j := 2*.}

h) In sections 2 up to 8 under "Semantics", a meaning is associated with occurrences of certain sequences of symbols by means of sentences in the English language, as a series of processes (the "elaboration" of those occurrences of sequences of symbols as terminal productions of given notions), each causing a specific effect. Any of these processes may be replaced by any process which causes the same effect.

i) If an occurrence of a sequence of symbols is the offspring of a direct descendent *D* of a notion *N* which is the only member of a direct production of *N*, then its "preelaboration", possibly yielding a "prevalue" with a "premode" and a "prescope", as terminal production of *N* is its elaboration, possibly yielding a "value" with a "mode" and a "scope", as terminal production of *D* and, except as otherwise specified {8.2}, its elaboration with value, mode and scope as terminal production of *N* is its preelaboration with prevalue, premode and prescope as terminal production of *N*.

{The elaboration with value, mode and scope of the reference-to-real-confrontation (8.3.0.1.a) *x := 3.14* is its preelaboration with prevalue, premode and prescope which is its elaboration with value, mode and scope as a reference-to-real-assignation.

The syntax of the strict language has been chosen in such a way that a given sequence of symbols which is a terminal production of 'program' is so by means of a unique set of productions, except, possibly, for production rules inducing at most preelaboration, e.g., derived from rules 6.1.1.g, 6.2.1.e and 6.4.1.d (balancing of modes) and from rule 7.1.1.cc in combination with 7.2.1.a and 7.4.1.a (order of terminal productions of 'MOOD' in a terminal production of 'UNITED'); see also 2.3.a.}

j) A terminal production of a metanotion *M* is "enveloped" by a notion *N* if it is contained once in *N* but not in another terminal production of *M* contained in *N*.

{Thus, 'reference to real' is enveloped as terminal production of 'MODE' by 'reference to real mode identifier', but 'real' is not.}

k) If something is left "undefined" or is said to be undefined, then this means that it is not defined by this Report alone, and that, for its definition, information from outside this Report has to be taken into account.

1.1.7. The Extended Language

The extended language encompasses the strict language; i.e., a program in the strict language, possibly subjected to a number of notational changes by virtue of "extensions" given in Chapter 9, is a program in the extended language and has the same meaning.

{Thus, **real** x is a representation of an identity-declaration in the extended language which means the same as **ref real** $x =$ **loc real** which is a representation of that identity-declaration in the strict language; see 9.2.a.}

1.1.8. The Representation Language

a) The representation language represents the extended language; i.e., a program in the extended language, in which all symbols are replaced by certain typographical marks by virtue of "representations", given in section 3.1.1, and in which all commas {not comma-symbols} are deleted, is a program in the representation language and has the same meaning.

b) Each version of the language in which representations are used which are sufficiently close to the given representations to be recognized without further elucidation is also a representation language. A version of the language in which notations or representations are used which are not obviously associated with those defined here, is a "publication language" or "hardware language" {, i.e., a version of the language suited to the supposed preference of the human or mechanical interpreter of the language}.

{e.g., *begin*, **begin**, 'BEGIN and 'BEGIN' (3.1.2.b) are all representations of the begin-symbol (3.1.1.e) in the representation language and some combination of holes in a punched card may be a representation of it in some hardware language.}

1.2. The Metaproduction Rules

1.2.1. Metaproduction Rules of Modes

a) MODE : MOOD ; UNITED.
b) MOOD : TYPE ; STOWED.
c) TYPE : PLAIN ; format ; PROCEDURE ; reference to MODE.
d) PLAIN : INTREAL ; boolean ; character.
e) INTREAL : INTEGRAL ; REAL.
f) INTEGRAL : LONGSETY integral.

g) REAL : LONGSETY real.
h) LONGSETY : long LONGSETY ; EMPTY.
i) EMPTY : .
j) PROCEDURE : procedure PARAMETY MOID.
k) PARAMETY : with PARAMETERS ; EMPTY.
l) PARAMETERS : PARAMETER ; PARAMETERS and PARAMETER.
m) PARAMETER : MODE parameter.
n) MOID : MODE ; void.
o) STOWED : structured with FIELDS ; row of MODE.
p) FIELDS : FIELD ; FIELDS and FIELD.
q) FIELD : MODE field TAG.
r) TAG : LETTER ; TAG LETTER ; TAG DIGIT.
s) LETTER : letter ALPHA ; letter aleph.
t) ALPHA : a ; b ; c ; d ; e ; f ; g ; h ; i ; j ; k ; l ; m ; n ; o ; p ; q ; r ; s ; t ; u ; v ; w ; x ; y ; z.
u) DIGIT : digit FIGURE.
v) FIGURE : zero ; one ; two ; three ; four ; five ; six ; seven ; eight ; nine.
w) UNITED : union of LMOODS MOOD mode.
x) LMOODS : LMOOD ; LMOODS LMOOD.
y) LMOOD : MOOD and.

{The reader might find it helpful to note that a metanotion ending with 'ETY' always has 'EMPTY' as a production.}

1.2.2. Metaproduction Rules Associated with Modes

a) PRIMITIVE : integral ; real ; boolean ; character ; format.
b) ROWS : row of ; ROWS row of.
c) ROWSETY : ROWS ; EMPTY.
d) ROWWSETY : ROWSETY.
e) NONROW : NONSTOWED ; structured with FIELDS.
f) NONSTOWED : TYPE ; UNITED.
g) REFETY : reference to ; EMPTY.
h) NONPROC : PLAIN ; format ; procedure with PARAMETERS MOID ; reference to NONPROC ; structured with FIELDS ; row of NONPROC ; UNITED.
i) PRAM : procedure with LMODE parameter and RMODE parameter MOID ; procedure with RMODE parameter MOID.
j) LMODE : MODE.
k) RMODE : MODE.
l) LMOOT : MOOD and.
m) LMOODSETY : MOOD and LMOODSETY ; EMPTY.
n) RMOODSETY : RMOODSETY and MOOD ; EMPTY.
o) LOSETY : LMOODSETY.
p) BOX : LMOODSETY box.
q) LFIELDSETY : FIELDS and ; EMPTY.
r) RFIELDSETY : and FIELDS ; EMPTY.
s) COMPLEX : structured with real field letter r letter e and real field letter i letter m.

t) BITS : structured with row of boolean field LENGTHETY letter aleph.
u) LENGTHETY : LENGTH LENGTHETY ; EMPTY.
v) LENGTH : letter l letter o letter n letter g.
w) BYTES : structured with row of character field LENGTHETY letter aleph.
x) STRING : row of character ; character.
y) MABEL : MODE mode ; label.

1.2.3. Metaproduction Rules Associated with Phrases and Coercion

a) PHRASE : declaration ; CLAUSE.
b) CLAUSE : MOID clause.
c) SOME : serial ; unitary ; CLOSED ; choice ; THELSE.
d) CLOSED : closed ; collateral ; conditional.
e) THELSE : then ; else.
f) SORTETY : SORT ; EMPTY.
g) SORT : strong ; FEAT.
h) FEAT : firm ; weak ; soft.
i) STRONGETY : strong ; EMPTY.
j) STIRM : strong ; firm.
k) ADAPTED : ADJUSTED ; widened ; rowed ; hipped ; voided.
l) ADJUSTED : FITTED ; procedured ; united.
m) FITTED : dereferenced ; deprocedured.

1.2.4. Metaproduction Rules Associated with Coercends

a) COERCEND : MOID FORM.
b) FORM : confrontation ; FORESE.
c) FORESE : ADIC formula ; cohesion ; base.
d) ADIC : PRIORITY ; monadic.
e) PRIORITY : priority NUMBER.
f) NUMBER : one ; TWO ; THREE ; FOUR ; FIVE ; SIX ; SEVEN ; EIGHT ; NINE.
g) TWO : one plus one.
h) THREE : TWO plus one.
i) FOUR : THREE plus one.
j) FIVE : FOUR plus one.
k) SIX : FIVE plus one.
l) SEVEN : SIX plus one.
m) EIGHT : SEVEN plus one.
n) NINE : EIGHT plus one.

1.2.5. Other Metaproduction Rules

a) VICTAL : VIRACT ; formal.
b) VIRACT : virtual ; actual.
c) LOWPER : lower ; upper.
d) ANY : KIND ; suppressible KIND ; replicatable KIND ; replicatable suppressible KIND.
e) KIND : sign ; zero ; digit ; point ; exponent ; complex ; string ; character.
f) NOTION : ALPHA ; NOTION ALPHA.

g) SEPARATOR : LIST separator ; go on symbol ; completer ; sequencer.
h) LIST : list ; sequence.
i) PACK : pack ; package.

{Rule f implies that all protonotions (1.1.2.b) are productions (1.1.3.e) of the metanotion (1.1.3.b) 'NOTION'; for the use of this metanotion see 3.0.1.b,c,d, g,h,i.}

{"*Well, 'slithy' means 'lithe and slimy'.... You see it's like a portmanteau — there are two meanings packed up into one word.*"
Through the Looking-glass, Lewis Carroll.}

1.3. Pragmatics

Scattered throughout this Report are "pragmatic" remarks included between the braces { and }. These do not form part of the definition of the language but are intended to help the reader to understand the implications of the definitions and to find corresponding sections or rules.

{The rules under Syntax are provided with cross-references to be interpreted as follows. Let a "hypernotion" be either a protonotion or a sequence of one or more metanotions, possibly preceded and/or separated and/or followed by protonotions; then each rule consists of a hypernotion followed by a colon followed by one or more hypernotions separated by commas or semicolons, and is closed by a point. By virtue of 1.1.5.a.Step 3, each hypernotion eventually yields one or more protonotions. In each rule, a hypernotion before (after) the colon is followed by indicators of the rules in which a hypernotion yielding one or more protonotions also yielded by the first hypernotion appears after (before) the colon, or by indicators of the representations in section 3.1.1 of the symbols yielded by the first hypernotion. Here, an indicator is, in principle, the section number followed by the letter indicating the line where the rule or representation appears, with the following conventions:

i) the indicators whose section number is that of the section in which they appear, are given first and their section number is omitted; e.g., "3.0.3.b" appears as "b" in section "3.0.3";

ii) all points are omitted and 10 appears as A; e.g., "3.0.3.a" appears as "303a" elsewhere and "3.0.10.a" appears as "30Aa";

iii) a final 1 is omitted; e.g., "811a" appears as "81a";

iv) a section number which is the same as in the preceding indicator is omitted; e.g., "821a,821b" appears as "821a,b";

v) numerous indicators of the rules 3.0.1.b up to i are replaced by more helpful indicators; e.g., in 6.1.1.c, "chain of strong void units separated by go on symbols{30c}" appears as "chain of strong void units{e} separated by go on symbols{31f}"; also, indicators in Section 3.0.1 are restricted to a bare minimum;

vi) the absence of a production rule for one or more protonotions which are not symbols and are yielded by the hypernotion appearing after the colon, is indicated by "—"; e.g., in 8.3.0.1.a after "MODE conformity relation" appears

{832a,—} since 8.3.2.1.a yields only production rules for "boolean conformity relation", and no other rule provides the absent productions.}

{Some of the pragmatic remarks are examples in the representation language. In these examples, identifiers occur out of context from their defining occurrences. Unless otherwise specified, these occurrences identify those in the standard-prelude (2.1.b and Chapter 10) (e.g., see 10.3.k for *random* and 10.3.a for *pi*), that in the exit (2.1.e) (viz., *exit*), or those in:

int *i, j, k, m, n;* **real** *a, b, x, y;* **bool** *p, q, overflow;* **char** *c,* **format** *f;* **bytes** *r;* **string** *s;* **bits** *t;* **compl** *w, z;* **ref real** *xx, yy;* *[1:n]* **real** *x1, y1;* *[1:m, 1:n]* **real** *x2;* *[1:n, 1:n]* **real** *y2;* *[1:n]* **int** *i1;* *[1:m, 1:n]* **int** *i2;*
proc ref real *x or y* = **ref real** : *(random < .5 | x | y);*
proc *ncos* = *(***int** *i)* **real** : *cos (2×pi×i / n);*
proc *nsin* = *(***int** *i)* **real** : *sin (2×pi×i / n);*
proc *g* = *(***real** *u)* **real** : *(arctan (u) − a + u − 1);*
proc *stop* = **go to** *exit;*
princeton: grenoble: st pierre de chartreuse: kootwijk: warsaw: zandvoort: amsterdam: tirrenia: north berwick: munich: stop .}

{Merely corroborative detail, intended to give artistic verisimilitude to an otherwise bald and unconvincing narrative.
Mikado, W. S. Gilbert.}

2. The Computer and the Program

{The programmer is concerned with particular-programs (2.1.d). These are always contained in a program (2.1.a), which also contains the standard-prelude, i.e., a declaration-prelude which is always the same (see Chapter 10), possibly a library-prelude, i.e., a declaration-prelude which may depend upon the implementation, the exit, i.e., *; exit:* , which enables the programmer to end the elaboration of a program anywhere by that of the jump *exit*, possibly a library-postlude, and the standard-postlude (10.6).}

2.1. Syntax

a) program : open symbol{31e}, standard prelude{b}, library prelude{c} option, particular program{d}, exit{e}, library postlude{f} option, standard postlude{g}, close symbol{31e}.

b) standard prelude{a} : declaration prelude{61b} sequence.

c) library prelude{a} : declaration prelude{61b} sequence.

d) particular program{a} : label{61k} sequence option, strong CLOSED void clause{62b,63a,64a}.

e) exit{a} : go on symbol {31f}, letter e letter x letter i letter t{41c}, label symbol{31e}.

f) library postlude{a} : statement interlude{61i}.

g) standard postlude{a} : strong void clause train{61h}.

2.2. Terminology

{"*When I use a word,*" *Humpty Dumpty said, in rather a scornful tone,* "*it means just what I choose it to mean — neither more nor less.*"
Through the Looking-glass, Lewis Carroll.}

The meaning of a program is explained in terms of a hypothetical computer which performs a set of "actions" {2.2.5}, the elaboration of the program {2.3.a}. The computer deals with a set of "objects" {2.2.1} between which, at any given time, certain "relationships" {2.2.2} may "hold".

2.2.1. Objects

Each object is either "external" or "internal". External objects are occurrences of symbols or of terminal productions {1.1.2.f} of notions. Internal objects are "instances" of values {2.2.3}.

2.2.2. Relationships

a) Relationships either are "permanent", i.e., independent of the program and its elaboration, or actions may cause them to hold or cease to hold. Each relationship is either between external objects or between an external object and an internal object or between internal objects.

b) The relationships between external objects are: to contain {1.1.6.b}, to be a direct constituent of {1.1.6.e} and "to identify" {c}.

c) A given occurrence of a terminal production of 'MABEL identifier' {4.1.1.b} ('MODE mode indication' {4.2.1.b} or 'PRIORITY indication' {4.2.1.e}, 'PRAM ADIC operator' {4.3.1.b,c}) where "MABEL" ("MODE", "PRIORITY", "PRAM", "ADIC") stands for any terminal production of the metanotion 'MABEL' ('MODE', 'PRIORITY', 'PRAM', 'ADIC') may identify a "defining occurrence" ("indication-defining occurrence", "operator-defining occurrence") of the same terminal production.

d) The relationship between an external object and an internal object is: "to possess".

e) An external object considered as an occurrence of a terminal production of a given notion may possess an instance of a value, termed "the" value of the external object when it is clear which notion is meant; in general, "an (the) instance of a (the) value" is sometimes shortened in the sequel to "a (the) value" when it is clear which instance is meant

f) A mode-identifier (operator) may possess a value ({more specifically} a "routine" {2.2.3.4}). This relationship is caused to hold by the elaboration of an identity-declaration {7.4.1.a} (operation-declaration {7.5.1.a}) and ceases to hold upon the end of the elaboration of the smallest serial-clause {6.1.1.a} containing that declaration.

g) An external object other than an identifier or operator {e.g., serial-clause (6.1.1.a)} considered as a terminal production of a given notion may be caused to possess a value by its elaboration as terminal production of that notion, and continues to possess that value until the next elaboration, if any, of that external object is "initiated", whereupon it ceases to possess that value.

h) The relationships between internal objects are: "to be of the same mode as", "to be equivalent to", "to be smaller than", "to be a component of" and "to refer to". A relationship said to hold between a given value and a (an instance of a) second value holds between any instance of the given value and any (that) instance of the second value.

i) An instance of a value may be of the same mode as another one; this relationship is permanent {2.2.4.1.a}.

j) A value may be equivalent to another value {2.2.3.1.d,f} and a value may be smaller than another value {2.2.3.1.c}. If one of these relationships is defined at all for a given pair of values, then either it does not hold, or it does hold and is permanent.

k) An instance of a given value is a component of another one if it is a "field" {2.2.3.2}, "element" {2.2.3.3.a} or "subvalue" {2.2.3.3.c} of that other one or of one of its components.

l) Any "name" {2.2.3.5}, except "*nil*" {2.2.3.5.a}, refers to one instance of another value. This relationship {may be caused to hold by an "assignment" (8.3.1.2.c) of that instance of that value to that name and} continues to hold until another instance of a value is caused to be referred to by that name. The words "refers to an instance of" are often shortened in the sequel to "refers to".

2.2.3. Values

Values are "plain values" {2.2.3.1}, "structured values" {2.2.3.2}, "multiple values" {2.2.3.3}, routines {2.2.3.4}, "formats" {2.2.3.4}, and names {2.2.2.1, 2.2.3.5}.

2.2.3.1. Plain Values

a) A plain value is either an "arithmetic value", i.e., an "integer" or a "real number", or is a "truth value" or a "character".

b) An arithmetic value has a "length number", i.e., a positive integer characterizing the degree of discrimination with which the value is kept in the computer. The number of integers (real numbers) of given length number that can be distinguished increases with the length number up to a certain length number, the number of different lengths of integers (real numbers) {10.1.a,c}, after which it is constant.

c) For each pair of integers (real numbers) of the same length number, the relationship "to be smaller than" is defined {10.2.3.a, 10.2.4.a}. For each pair of integers of the same length number, a third integer of that length number may exist, the first integer "minus" the other one {10.2.3.g}. Finally, for each pair of real numbers of the same length number, three real numbers of that length number may exist, the first real number "minus" ("times", "divided by") the other one {10.2.4.g,l,m}; these real numbers are obtained "in the sense of numerical analysis", i.e., by performing the operations known in mathematics by these terms on real numbers which may deviate slightly from the given ones {; this deviation is left undefined in this Report}.

d) Each integer of given length number is equivalent to a real number of that length number. Also, each integer (real number) of given length number is equivalent to an integer (real number) whose length number is greater by one. These equivalences permit the "widening" {8.2.5} of an integer into a real number and the increase of the length number of an integer or real number {10.2.3.q, 10.2.4.n}. The inverse transformations are only possible on those real numbers (arithmetic values) which are equivalent to an integer {10.2.4.p} (a value of smaller length number {10.2.3.r, 10.2.4.o}).

e) A truth value is either "*true*" or "*false*".

f) Each character is equivalent to a nonnegative integer of length number one, its "integral equivalent" {10.1.j}; this relationship is defined only to the extent that different characters have different integral equivalents.

2.2.3.2. Structured Values

A structured value is composed of a number of other values, its fields, in a given order, each of which is "selected" {8.5.2.2.Step 2} by a specific field-selector {7.1.1.i}.

2.2.3.3. Multiple Values

a) A multiple value is composed of a "descriptor" and a number of other values, its elements, each of which is selected {8.6.1.2.Step 7} by a specific integer, its "index".

b) The descriptor consists of an "offset", c, and some number, $n > 0$, of "quintuples $(l_i, u_i, d_i, s_i, t_i)$ of integers, $i = 1, \ldots, n$; l_i is the i-th "lower bound", u_i the i-th "upper bound", d_i the i-th "stride", s_i the i-th "lower state" and t_i the i-th "upper state". If for any i, $i = 1, \ldots, n$, $u_i < l_i$, then the number of elements in the multiple value is zero; otherwise, it is $(u_1 - l_1 + 1) \times \ldots \times (u_n - l_n + 1)$. The descriptor "describes" each element selected by $c + (r_1 - l_1) \times d_1 + \ldots + (r_n - l_n) \times d_n$ where $l_i \leq r_i \leq u_i$ for each $i = 1, \ldots, n$.

{To the name referring to a given multiple value a state of which is 1, no multiple value can be assigned (8.3.1.2.c.Step 4) in which the bound corresponding to that state differs from that in the given value.}

c) A subvalue of a given multiple value is a multiple value which is referred to by the value of a slice {8.6.1.1.a} the value of whose primary refers to the given multiple value.

d) If each element of an instance I of a multiple value M is the same instance of a value as the corresponding element of an instance J of M, then I and J are one same instance of M.

2.2.3.4. Routines and Formats

A routine (format) is a sequence of symbols which is the same as some closed-clause {6.3.1.a} (format-denotation {5.5.1.a}).

2.2.3.5. Names

a) There is one name, *nil*, whose scope {2.2.4.2} is the program and which does not refer to any value; any other name is created by the elaboration of an actual-declarer {7.1.2.d.Step 8}, a rowed-coercend {8.2.6.2.Step 7} or a skip {8.2.7.2.a} {, and refers to precisely one instance of a value}.

b) If a given name *N* refers to a structured value {2.2.3.2}, then to each of its fields there refers a name uniquely determined by *N* and the field-selector selecting that field, and whose scope is that of *N*.

c) If a given name *N* refers to a multiple value *M* {2.2.3.3}, then to each element of *M* (each multiple value composed of a descriptor and elements which are a proper subset of the elements of *M* or composed of a descriptor different from that of *M* and the elements of *M*) there refers a name uniquely determined by *N* and the index of that element (and that descriptor and those elements), and whose scope is that of *N*.

d) If an instance of a name and an instance of a second name refer to one same instance of a value, then they are instances of one same name.

2.2.4. Modes and Scopes

2.2.4.1. Modes

a) A mode is any terminal production of "MODE' {1.2.1.a}. Each instance of a value is of one specific mode which is a terminal production of 'MOOD' {1.2.1.b}; furthermore, all instances of a given value *V* other than *nil* {2.2.3.5.a} are of one same mode, the mode of *V*, and a "copy" of a given instance *I* of a value *V* is a new instance of *V* which is of the same mode as *I*.

b) The mode of a truth value (character, format) is 'boolean' ('character', 'format').

c) The mode of an integer (a real number) of length number *n* is $(n-1)$ times 'long' followed by 'integral' (by 'real').

d) The mode of a structured value is 'structured with' followed by one or more "portrayals" separated by 'and', one corresponding to each field taken in the same order, each portrayal being the mode of that field followed by 'field' followed by a terminal production of 'TAG' {1.2.1.r} whose terminal production {field-selector} selects {2.2.3.2} that field; it is "structured from" a second mode if the mode in one of its portrayals is, or is structured from, it.

e) The mode of a multiple value is a terminal production of 'NONROW' {1.2.2.e} preceded by as many times 'row of' as there are quintuples in the descriptor of that value.

f) The mode of a routine is a terminal production of 'PROCEDURE' {1.2.1.j}.

g) The mode of a name is 'reference to' followed by another mode; if the name is not *nil*, then that other mode is either the mode of the value to which the name refers, or is united from {4.4.3.a} it. {See 7.1.2.d.Step 8.}

2.2.4.2. Scopes

a) Each value has one specific scope.

b) The scope of a plain value is the program,
that of a structured (multiple) value is the smallest of the scopes of its fields (elements, if any, and, otherwise, the program),
that of a routine or format possessed by a given denotation *D* {5.4.1.a, 5.5.1.a} (of a routine which is the same sequence of symbols as a given cast-pack *D* or void-cast-pack-pack *D* {8.3.4.1.a}) is the smallest range {4.1.1.e} containing a defining occurrence {4.1.2.a} (indication-defining occurrence {4.2.2.a}, operator-defining occurrence {4.3.2.a}) of a terminal production of a notion ending with 'identifier' ('indication', 'operator'), if any, not contained in *D* but identified {4.1.2.b (4.2.2.b, 4.3.2.b)} by some applied (indication-applied, operator-applied) occurrence of that terminal production contained in *D*, and, otherwise, the program, and
that of a name is some {2.2.3.5, 8.5.1.2.b} range.

2.2.5. Actions

{*Suit the action to the word, the word to the action.*
Hamlet, William Shakespeare.}

a) An action is "inseparable", "serial" or "collateral". A serial action consists of actions which take place one after the other.

b) A collateral action consists of actions merged in time; i.e., it consists of inseparable actions each of which is chosen, in a way which is left undefined in this Report, from amongst the first of the inseparable actions which, at that moment, according to this Report, would be the continuation of any of the constituting actions.

c) The elaboration of any (of the closed-clause following the do-symbol {3.1.1.h} in any) closed-clause {6.3.1.a} which is a modified copy {8.4.2} of the actual-parameter of the operation-declaration {7.5.1.a} 10.4.d (10.4.c) is an inseparable action.

{What other actions are inseparable is left undefined in this Report.}

2.3. Semantics

{*"I can explain all the poems that ever were invented — and a good many that haven't been invented just yet."*
Through the Looking-glass, Lewis Carroll.}

a) The elaboration of a program is the elaboration of the strong-closed-void-clause {6.3.1.a} consisting of the same sequence of symbols.

{In this Report, the syntax says which sequences of symbols are programs and the semantics which actions are performed by the computer when elaborating a program. Both syntax and semantics are recursive. Though certain sequences of symbols may be terminal productions of 'program' in more than one way (1.1.6.i), this syntactic ambiguity does not lead to a semantic ambiguity.}

b) In ALGOL 68, a specific notation for external objects is used which, together with its recursive definition, makes it possible to handle and to distinguish between arbitrarily long sequences of symbols, to distinguish between arbitrarily many different values of a given mode (except 'boolean') and to distinguish between arbitrarily many modes, which allows arbitrarily many objects to exist within the computer and which allows the elaboration of a program to involve an arbitrarily large, not necessarily finite, number of actions. This is not meant to imply that the notation of the objects in the computer is that used in ALGOL 68 nor that it has the same possibilities. It is, on the contrary, not assumed that the computer can handle arbitrary amounts of presented information. It is not assumed that these two notations are the same or even that a one-to-one correspondence exists between them; in fact, the set of different notations of objects of a given category may be finite. It is not assumed that the speed of the computer is sufficient to elaborate a given program within a prescribed lapse of time, nor that the number of objects and relationships that can be established is sufficient to elaborate it at all.

c) A model of the hypothetical computer, using a physical machine, is said to be an "implementation" of ALGOL 68, if it does not restrict the use of the language in other respects than those mentioned above. Furthermore, if a language is defined whose particular-programs are particular-programs of ALGOL 68 and have the same meaning, then that language is said to be a sublanguage of ALGOL 68. A model is said to be an implementation of a sublanguage if it does not restrict the use of the sublanguage in other respects than those mentioned above.

{A sequence of symbols which is not a program but can be turned into one by deleting or inserting a certain number of symbols and not a smaller number could be regarded as a program with that number of syntactical errors. Any program that can be obtained by deleting or inserting that number of symbols may be termed a "possibly intended" program. Whether a program or one of the possibly intended programs has the effect its author in fact intended it to have, is a matter which falls outside this Report.}

{In an implementation, the particular-program may be "compiled", i.e., translated into an "object program" in the code of the physical machine. Under certain circumstances, it may be advantageous to compile parts of the particular-program independently, e.g., parts which are common to several particular-programs. If such a part contains mode-identifiers (indications, operators) whose defining (indication-defining, operator-defining) occurrences (Chapter 4) are not contained in that part, then compilation into an efficient object program may be assured by preceding the part by a chain of formal-parameters (5.4.1.e) (mode-declarations (7.2.1.a) or priority-declarations (7.3.1.a), captions (7.5.1.b)) containing those defining (indication-defining, operator-defining) occurrences.}

{The definition of specific sublanguages and also the specification of actions not definable by any program (e.g., compilation or initiation of the elaboration), is not given in this Report. However, the definition of the language allows, for instance, to let a special representation of the comment-symbol different from the ones given in 3.1.1.i, viz., ¢, #, **co** or **comment**, preferably **pr** or **pragmat**, have the effect that by a comment (3.0.9.b) beginning and ending with this special representation, the computer is invited to implement some such sublanguage or ALGOL 68 itself or to take some such undefinable action, as may be specified by the comment (e.g., **pr** *algol 68* **pr**, **pr** *run* **pr** or **pr** *dump* **pr**).}

{**pr** *algol 68* **pr**
begin
proc pr *nonrec* **pr** *p* = (: *p*);
p
end
pr *run* **pr pr** ? **pr**
Report on the Algorithmic
Language ALGOL 68.}

3. Basic Tokens and General Constructions

3.0. Syntax

3.0.1. Introduction

a)⋆ basic token : letter token{302a} ; denotation token{303a} ;
action token{304a} ; declaration token{305a} ;
syntactic token{306a} ; sequencing token{307a} ;
hip token{308a} ; extra token{309a} ; special token{30Aa}.
b) NOTION option : NOTION ; EMPTY.
c) chain of NOTIONs separated by SEPARATORs{c,d} : NOTION ;
NOTION, SEPARATOR{e,f,31f,61j,l},
chain of NOTIONs separated by SEPARATORs{c}.
d) NOTION LIST{g} : chain of NOTIONs separated by LIST separators{c,e,f}.
e) list separator{c,d,g} : comma symbol{31e}.
f) sequence separator{c,d,g} : EMPTY.
g) NOTION LIST proper : NOTION, LIST separator{e,f}, NOTION LIST{d}.
h) NOTION pack : open symbol{31e}, NOTION, close symbol{31e}.
i) NOTION package : begin symbol{31e}, NOTION, end symbol{31e}.

{Examples:
a) *a* ; *0* ; + ; **int** ; **if** ; . ; **nil** ; **for** ; " ;
b) *0* ; ; (integral-part-options)
c) *0, 1, 2* ; (a chain-of-strong-integral-units-separated-by-list-separators)
d) *0* ; *0, 1, 2* ; (strong-integral-unit-lists)
e) , ;
g) *1, 2, 3* ; (a strong-integral-unit-list-proper)
h) *(1, 2, 3)* ; (a strong-integral-unit-list-proper-pack)
i) **begin** *x* := *3.14*; *y* := *2.72* **end** (a strong-serial-void-clause-package) }

3.0.2. Letter Tokens

a)⋆ letter token : LETTER{b}.
b) LETTER{309d,41b,c,d,512h,55h,i,o,q,552b,e,f,553f,554a,555b,556c,557c,71j} :
LETTER symbol{31a}.

{Examples:
a) *a* ; (see 1.1.4.Step 2)}

{Letter-tokens either are, or are constituents of, identifiers (4.1.1.a), field-selectors (7.1.1.i), real-denotations (5.1.2.1.a), format-denotations (5.5.1.a) and string-items (5.3.1.d).}

3.0.3. Denotation Tokens

a)⋆ denotation token : number token{b} ; true symbol{31b} ;
  false symbol{31b} ; formatter symbol{31b} ; flipflop{e} ; space symbol{31b}.
b) number token{309d} : digit token{c} ; point symbol{31b} ;
  times ten to the power symbol{31b}.
c) digit token{b,511a} : DIGIT{d}.
d) DIGIT{c,41d,552c} : DIGIT symbol{31b}.
e) flipflop{309d,52c} : flip symbol{31b} ; flop symbol{31b}.

{Examples:
a) *1* ; **true** ; **false** ; $; **1** ; **.** ;
b) *1* ; . ; $_{10}$;
c) *1* ;
d) *1* ;
e) **1** ; **0** }

{Denotation-tokens are, or are constituents of, denotations (5.0.1.a). Some denotation-tokens may, by themselves, be denotations, e.g., the digit-token *1*, whereas others, e.g., the formatter-symbol $, serve only to construct denotations.}

3.0.4. Action Tokens

a)⋆ action token : operator token{b} ; equals symbol{31c} ;
  times symbol{31c} ; confrontation token{d}.
b) operator token{42e,f} : minus and becomes symbol{31c} ;
  plus and becomes symbol{31c} ; times and becomes symbol{31c};
  divided by and becomes symbol{31c} ; over and becomes symbol{31c} ;
  modulo and becomes symbol{31c} ; prus and becomes symbol{31c} ;
  or symbol{31c} ; and symbol{31c} ; differs from symbol{31c} ;
  is less than symbol{31c} ; is at most symbol{31c} ;
  is at least symbol{31c} ; is greater than symbol{31c} ;
  plusminus{c} ; divided by symbol{31c} ; over symbol{31c} ;
  modulo symbol{31c} ; th element of symbol{31c} ;
  lower bound of symbol{31c} ; upper bound of symbol{31c} ;
  lower state of symbol{31c} ; upper state of symbol{31c} ;
  plus i times symbol{31c} ; not symbol{31c} ; down symbol{31c} ;
  up symbol{31c} ; absolute value of symbol{31c} ; binal symbol{31c} ;
  representation of symbol{31c} ; lengthen symbol{31c} ;
  shorten symbol{31c} ; odd symbol{31c} ; sign symbol{31c} ;
  round symbol{31c} ; real part of symbol{31c} ;
  imaginary part of symbol{31c} ; conjugate of symbol{31c} ;
  booleans to bits symbol{31c} ; characters to bytes symbol{31c}.
c) plusminus{b,512i,55p} : plus symbol{31c} ; minus symbol{31c}.

d)⋆ confrontation token : becomes symbol{31c} ; conforms to symbol{31c} ;
conforms to and becomes symbol{31c} ; is symbol{31c} ;
is not symbol{31c} ; cast of symbol{31c}.

{Examples:

a) + ; = ; × ; := ;

b) −:= ; +:= ; ×:= ; /:= ; ÷:= ; ÷::= ; +=: ; ∨ ; ∧ ; ≠ ; < ; ≤ ; ≥ ; > ;
+ ; / ; ÷ ; ÷: ; ▯ ; ⌊ ; ⌈ ; ⌊ ; ⌈ ; ⊥ ; ¬ ; ↓ ; ↑ ; **abs** ; **bin** ; **repr** ; **leng** ; **short** ;
odd ; **sign** ; **round** ; **re** ; **im** ; **conj** ; **btb** ; **ctb** ;

c) + ; − ;

d) := ; :: ; ::= ; :=: ; :≠: ; :}

{Operator-tokens are constituents of formulas (8.4.1.a). Confrontation-tokens are constituents of confrontations (8.3.0.1.a).}

3.0.5. Declaration Tokens

a)⋆ declaration token : PRIMITIVE symbol{31d} ; long symbol{31d} ;
structure symbol{31d} ; reference to symbol{31d} ;
flexible symbol{31d} ; either symbol{31d} ; procedure symbol{31d} ;
union of symbol{31d} ; mode symbol{31d} ; complex symbol{31d} ;
bits symbol{31d} ; bytes symbol{31d} ; string symbol{31d} ;
sema symbol{31d} ; file symbol{31d} ; priority symbol{31d} ;
local symbol{31d} ; heap symbol{31d} ; operation symbol{31d}.

{Examples:

a) **int** ; **long** ; **struct** ; **ref** ; **flex** ; **either** ; **proc** ; **union** ; **mode** ; **compl** ; **bits** ;
bytes ; **string** ; **sema** ; **file** ; **priority** ; **loc** ; **heap** ; **op** }

{Declaration-tokens either are, or are constituents of, declarers (7.1.1.a), which specify modes (2.2.4), or of declarations (7.2.1.a, 7.3.1.a, 7.4.1.a, 7.5.1.a).}

3.0.6. Syntactic Tokens

a)⋆ syntactic token : open symbol{31e} ; close symbol{31e} ;
begin symbol{31e} ; end symbol{31e} ; comma symbol{31e} ;
parallel symbol{31e} ; sub symbol{31e} ; bus symbol{31e} ;
up to symbol{31e} ; at symbol{31e} ; if symbol{31e} ;
THELSE symbol{31e} ; fi symbol{31e} ; of symbol{31e} ;
label symbol{31e}.

{Examples:

a) (;) ; **begin** ; **end** ; , ; **par** ; [;] ; : ; @ ; **if** ; **then** ; **fi** ; **of** ; : }

{Syntactic-tokens separate external objects or group them together.}

3.0.7. Sequencing Tokens

a)⋆ sequencing token : go on symbol{31f} ; completion symbol{31f} ;
go to symbol{31f}.

{Examples:

a) ; ; . ; **go to** }

{Sequencing-tokens are constituents of clauses, in which they specify the order of elaboration (6.1.1.b,c,j,l, 8.2.7.1.c).}

3.0.8. Hip Tokens

a)⋆ hip token : skip symbol{31g} ; nil symbol{31g}.

{Examples:

a) **skip** ; **nil** }

{Hip-tokens function as skips and nihils (8.2.7.1.b,d).}

3.0.9. Extra Tokens and Comments

a)⋆ extra token : for symbol{31h} ; from symbol{31h} ; by symbol{31h} ; to symbol{31h} ; while symbol{31h} ; do symbol{31h} ; THELSE if symbol{31h}.

b) comment{9.1} : comment symbol{31i}, comment item{c} sequence option, comment symbol{31i}.

c) comment item{b} : character token{d} ; other comment item{1.1.5.c}.

d) character token{c,514b} : LETTER{302b} ; number token{303b} ; flipflop{303e} ; plus i times symbol{31c} ; open symbol{31e} ; close symbol{31e} ; comma symbol{31e} ; space symbol{31b}.

{Examples:

a) **for** ; **from** ; **by** ; **to** ; **while** ; **do** ; **thef** ;

b) # *with respect to* # ;

c) *w* ; ? ;

d) *a* ; *1* ; **1** ; **i** ; (;) ; , ; . }

3.0.10. Special Tokens

a)⋆ special token : quote symbol{31i} ; comment symbol{31i} ; indicant{1.1.5.b} ; dyadic indicant{1.1.5.b} ; monadic indicant{1.1.5.b}.

{Examples:

a) " ; # ; **primitive** ; ? ; ! }

3.1. Symbols

3.1.1. Representations

a) Letter tokens

symbol	representation	symbol	representation
letter a symbol{302b}	*a*	letter n symbol{302b}	*n*
letter b symbol{302b}	*b*	letter o symbol{302b}	*o*
letter c symbol{302b}	*c*	letter p symbol{302b}	*p*
letter d symbol{302b}	*d*	letter q symbol{302b}	*q*
letter e symbol{302b}	*e*	letter r symbol{302b}	*r*
letter f symbol{302b}	*f*	letter s symbol{302b}	*s*
letter g symbol{302b}	*g*	letter t symbol{302b}	*t*
letter h symbol{302b}	*h*	letter u symbol{302b}	*u*
letter i symbol{302b}	*i*	letter v symbol{302b}	*v*
letter j symbol{302b}	*j*	letter w symbol{302b}	*w*
letter k symbol{302b}	*k*	letter x symbol{302b}	*x*
letter l symbol{302b}	*l*	letter y symbol{302b}	*y*
letter m symbol{302b}	*m*	letter z symbol{302b}	*z*

{No representation for 'letter aleph symbol' is provided and the programmer cannot provide one himself; see 1.1.4.Step 2, 3.1.2.c.}

b) Denotation tokens

symbol	representation	
digit zero symbol{303d}	*0*	
digit one symbol{303d,73b}	*1*	
digit two symbol{303d,73c}	*2*	
digit three symbol{303d,73d}	*3*	
digit four symbol{303d,73e}	*4*	
digit five symbol{303d,73f}	*5*	
digit six symbol{303d,73g}	*6*	
digit seven symbol{303d,73h}	*7*	
digit eight symbol{303d,73i}	*8*	
digit nine symbol{303d,73j}	*9*	
point symbol{303b,512d,553c}		
times ten to the power symbol{303b,512h}	$_{10}$	\
true symbol{513a}	**true**	
false symbol{513a}	**false**	
formatter symbol{55a}	$	
flip symbol{303e}	**1**	
flop symbol{303e}	**0**	
space symbol{309d}	.	

c) Action tokens

symbol	representation		
minus and becomes symbol{304b}	−:=	**minus**	
plus and becomes symbol{304b}	+:=	**plus**	
times and becomes symbol{304b}	×:=	**times**	
divided by and becomes symbol{304b}	/:=	**div**	
over and becomes symbol{304b}	÷:=	**overb**	
modulo and becomes symbol{304b}	÷::=	**modb**	
prus and becomes symbol{304b}	+=:	**prus**	
or symbol{304b}	∨	**or**	
and symbol{304b}	∧	&	**and**
differs from symbol{304b}	≠	¬=	**ne**
is less than symbol{304b}	<		**lt**
is at most symbol{304b}	≤	<=	**le**
is at least symbol{304b}	≥	>=	**ge**
is greater than symbol{304b}	>		**gt**
divided by symbol{304b}	/		
over symbol{304b}	÷	**over**	
modulo symbol{304b}	÷:	**mod**	
th element of symbol{304b}	▯	**elem**	
lower bound of symbol{304b}	⌊	**lwb**	**entier**
upper bound of symbol{304b}	⌈	**upb**	
lower state of symbol{304b}	⌊	**lws**	
upper state of symbol{304b}	⌈	**ups**	

symbol	representation
plus i times symbol{304b,309d}	⊥ ! **i**
not symbol{304b}	¬ ~ **not**
down symbol{304b}	↓ **down**
up symbol{304b}	↑ ** ^ **up**
absolute value of symbol{304b}	**abs**
binal symbol{304b}	**bin**
representation of symbol{304b}	**repr**
lengthen symbol{304b}	**leng**
shorten symbol{304b}	**short**
odd symbol{304b}	**odd**
sign symbol{304b}	**sign**
round symbol{304b}	**round**
real part of symbol{304b}	**re**
imaginary part of symbol{304b}	**im**
conjugate of symbol{304b}	**conj**
booleans to bits symbol{304b}	**btb**
characters to bytes symbol{304b}	**ctb**
plus symbol{304c}	+
minus symbol{304c}	−
equals symbol{42e,72a,73a,74a,75a}	= **eq**
times symbol{42e}	× *
becomes symbol{831a}	:= ..= .=
conforms to symbol{832b}	:: **ct**
conforms to and becomes symbol{832b}	::= **ctab**
is symbol{833b}	:=: **is**
is not symbol{833b}	:≠: **isnt**
cast of symbol{834a}	: ..

d) Declaration tokens

symbol	representation
integral symbol{71c}	**int**
real symbol{71c}	**real**
boolean symbol{71c}	**bool**
character symbol{71c}	**char**
format symbol{71c}	**format**
long symbol{42c,e,f,510b,52b,71d}	**long**
structure symbol{71e}	**struct**
reference to symbol{71l,m,n}	**ref**
flexible symbol{71t,v}	**flex**
either symbol{71v}	**either**
procedure symbol{71w}	**proc**
union of symbol{71cc,ii}	**union**
mode symbol{72a}	**mode**
complex symbol{42c}	**compl**
bits symbol{42c}	**bits**
bytes symbol{42c}	**bytes**

symbol	representation		
string symbol{42c}	**string**		
sema symbol{42c}	**sema**		
file symbol{42c}	**file**		
priority symbol{73a}	**priority**		
local symbol{851b}	**loc**		
heap symbol{851c}	**heap**		
operation symbol{75b}	**op**		

e) Syntactic tokens

symbol	representation		
open symbol{2a,30h,309d,54b,554b}	(		
close symbol{2a,30h,309d,54b,554b}	)		
begin symbol{30i}	**begin**		
end symbol{30i}	**end**		
comma symbol{30e,309d,54d,554b,62e,g,71f,q,gg, 861b,c}	,		
parallel symbol{62b}	**par**		
sub symbol{71o,p,861a}	[		
bus symbol{71o,p,861a}	]		
up to symbol{71r,861f}	:	..	
at symbol{861g}	@	**at**	
if symbol{64a}	(	**if**	**case**
then symbol{64e}	|	**then**	**in**
else symbol{64e}	|	**else**	**out**
fi symbol{64a}	)	**fi**	**esac**
of symbol{852a}	→	**of**	
label symbol{2e,61k}	:	..	

f) Sequencing tokens

symbol	representation	
go on symbol{2e,30c,54d,61b,c,j}	;	.,
completion symbol{61l}	.	**exit**
go to symbol{827c}	**goto**	**go to**

g) Hip tokens

symbol	representation	
skip symbol{827b}	~	**skip**
nil symbol{827b}	∘	**nil**

h) Extra tokens

symbol	representation
for symbol{9.3.a,b}	**for**
from symbol{9.3.a,b,c}	**from**
by symbol{9.3.a,b,c}	**by**
to symbol{9.3.a,c}	**to**
while symbol{9.3.a,b,c}	**while**
do symbol{9.3.a,b,c}	**do**

symbol	representation
then if symbol{9.4.b}	|: **thef**
else if symbol{9.4.b,c}	|: **elsf**

i) Special tokens

symbol	representation
quote symbol{514a,c,53b}	" **quote**
comment symbol{309b}	¢ # **co** **comment**

3.1.2. Remarks

a) Where more than one representation of a symbol is given, any one of them may be chosen.

{However, discretion should be exercised, since the text

(a>b **then** *b | a* **esac** ,

though acceptable to an automaton, would be more intelligible to a human in either of the two representations

(a>b | b | a)

or

if *a>b* **then** *b* **else** *a* **fi** .

Also, some representations may not be available in a given implementation.}

b) A representation which is a sequence of underlined or bold-faced marks or a sequence of marks preceded by a "bold-face shift" {, e.g., apostrophe,} and ending on a "light-face shift" {, e.g., any mark different from the representation of a letter or digit,} or a sequence of marks between apostrophes is different from the sequence of those marks when not underlined, bold-faced, preceded by a bold-face shift and ending on a light-face shift or between apostrophes.

c) Representations of other terminal productions of 'letter token' {1.1.4.Step 2}, 'indicant', 'dyadic indicant', 'monadic indicant' {1.1.5.b}, 'other comment item' and 'other string item' {1.1.5.c} may be added, provided that no sequence of representations of symbols can be confused with any other such sequence.

{e.g., **do if** are representations of a do-symbol followed by an if-symbol, whereas **doif** might be an ill-chosen representation of an indicant.}

d) The fact that representations of the terminal productions of 'letter token' as given in 3.1.1.a are usually spoken of as small letters is not meant to imply that the so-called corresponding capital letters could not serve equally well as representations. On the other hand, if both a small letter and the corresponding capital letter occur, then one of them is the representation of another terminal production of 'letter token' {1.1.4.Step 2}.

{For certain different symbols, one same or nearly the same representation is given; e.g., for the cast-of-symbol, up-to-symbol and label-symbol respectively, the representations ":", ":" and ":", and, moreover, for all of them the representation ".." is given. It follows uniquely from the syntax which of these three symbols is represented by an occurrence of any mark similar to one of these representations outside comments and row-of-character-denotations. Also, some of the given representations appear to be "composite"; e.g., the representation

":=" of the becomes-symbol appears to consist of ":", the representation of the cast-of-symbol, etc., and "=", the representation of the equals-symbol, and the representation ".." of the cast-of-symbol, etc., appears to consist of "." and ".", each of which might be the representation of a point-symbol or completion symbol. It follows from the syntax that ":=" can occur outside comments and row-of-character-denotations as representation of the becomes-symbol only (since "=" cannot occur as representation of a monadic-operator). Similarly, the other given composite representations do not cause ambiguity.}

4. Identification and the Context Conditions

{A proper program is a program satisfying the context conditions, e.g., if *(**real** x; x := 1)* is contained in a proper program, then the second occurrence of *x* is a reference-to-real-mode-identifier not solely because of some production rule (though this might be possible with a more elaborate syntax) but also because it identifies the first occurrence according to one of the context conditions. This chapter describes the methods of identification and contains other context conditions which prevent such undesirable constructions as **mode a = a**.}

4.1. Identifiers

{Identifiers are sequences of letter-tokens and/or digit-tokens in which the first is a letter-token, e.g., *x1*. Mode-identifiers are made to possess values by the elaboration of identity-declarations (7.4.1.a). Some mode-identifiers possessing values which are not names might, in other languages, be termed constants, e.g., *m* in **int** *m = 4096*. Mode-identifiers possessing names which refer to such values might be termed variables and those possessing names which refer to names might be termed pointers. Such terminology is not used in this Report. Here, all mode-identifiers possess values, which are, or are not, names.}

4.1.1. Syntax

a)⋆ identifier : MABEL identifier{b}.
b) MABEL identifier{54e,61k,827c,860a} : TAG{c,d,302b}.
c) TAG LETTER{b,c,d,21e,71j} : TAG{c,d,302b}, LETTER{302b}.
d) TAG DIGIT{b,c,d,71j} : TAG{c,d,302b}, DIGIT{303d}.
e)⋆ range : program{2a}; SORTETY serial CLAUSE{61a}; procedure with PARAMETERS MOID denotation{54b}.

{Examples:
b) *x* ; *xx* ; *x1* ; *amsterdam*}

{Rule b together with 1.2.1.r and 1.2.2.y gives rise to an infinity of production rules of the strict language, one for each pair of terminal productions of 'MABEL' and 'TAG'. For example,

'real mode identifier : letter a letter b.'

is one such a production rule. From rules c and 3.0.2.b, one obtains

'letter a letter b : letter a, letter b.',
'letter a : letter a symbol.' and
'letter b : letter b symbol.',

yielding

'letter a symbol, letter b symbol'

as a terminal production of 'real mode identifier'. For additional insight into the function of rules c and d, see 7.1.1.j and 8.5.2.1.a.}

4.1.2. Identification of Identifiers

{The method of identification of identifiers is first to distinguish between defining and applied occurrences and then to discover which defining occurrence is identified by a given applied occurrence.}

a) A given occurrence of a terminal production of 'MABEL identifier' where "MABEL" stands for any terminal production of the metanotion 'MABEL' is a defining occurrence if it follows a formal-declarer {7.1.1.b}, or if it is contained in a label {6.1.1.k}; otherwise, it is an "applied occurrence".

b) If a given occurrence of a terminal production of 'MABEL identifier' (see a) is an applied occurrence, then it may identify a defining occurrence of the same terminal production found by the following steps:

Step 1: The given occurrence is termed the "home" and Step 2 is taken;

Step 2: If there exists a smallest range containing the home, then this range, with the exclusion of all ranges contained within it, is termed the home and Step 3 is taken {; otherwise, there is no defining occurrence which the given occurrence identifies; see 4.4.1.b};

Step 3: If the home contains a defining occurrence of the same terminal production of 'MABEL identifier', then the given occurrence identifies it; otherwise, Step 2 is taken.

{In the closed-clause (**string** *s* := "*abc*"; *s* [*3*] ≠ "*d*"), the first occurrence of *s* is a defining occurrence of a terminal production of 'reference to row of character mode identifier'. The second occurrence of *s* identifies the first and, in order to satisfy the identification condition (4.4.1.a), is also a terminal production of 'reference to row of character mode identifier'. Identifiers have no inherent meaning.}

4.2. Indications

{Indications are used for modes, priorities and operators. Some representations of indications chosen in this Report are sequences of bold-faced or underlined letters, e.g., **compl** and **plus**, but no production rule determines this sequence. The programmer may also create his own indications, provided that they cannot be confused with an other symbol (1.1.5.b, 3.1.2.c).}

4.2.1. Syntax

a)⋆ indication : MODE mode indication{b} ; ADIC indication{e,f}.

b) MODE mode indication{71b,ii,72a} : mode standard{c} ; indicant{1.1.5.b}.

c) mode standard{b} :
  string symbol{31d} ; sema symbol{31d} ; file symbol{31d} ;
  long symbol{31d} sequence option, complex symbol{31d} ;
  long symbol{31d} sequence option, bits symbol{31d} ;
  long symbol{31d} sequence option, bytes symbol{31d}.

d)⋆ dyadic indication : PRIORITY indication{e}.
e) PRIORITY indication{43b,73a} : dyadic indicant{1.1.5.b} ;
long symbol{31d} sequence option, operator token{304b} ;
long symbol{31d} sequence option, equals symbol{31c} ;
long symbol{31d} sequence option, times symbol{31c}.
f) monadic indication{43c} : monadic indicant{1.1.5.b} ;
long symbol{31d} sequence option, operator token{304b}.
g)⋆ adic indication : ADIC indication{e,f}.

{Examples:
b) **compl** ; **primitive** ;
c) **string** ; **sema** ; **file** ; **long compl** ; **bits** ; **long bytes** ;
e) ? ; + ; = ; × ;
f) / ; + ; **long btb** }

4.2.2. Identification of Indications

{The identification of indications is similar to that of identifiers.}

a) A given occurrence of a terminal production of 'MODE mode indication' ('PRIORITY indication') where "MODE" ("PRIORITY") stands for any terminal production of the metanotion 'MODE' ('PRIORITY') is an indication-defining occurrence if it precedes the equals-symbol of a mode-declaration {7.2.1.a} (priority-declaration {7.3.1.a}); otherwise, it is an "indication-applied occurrence".

b) If a given occurrence of a terminal production of "MODE mode indication' ('PRIORITY indication') (see a) is an indication-applied occurrence, then it may identify an indication-defining occurrence of the same terminal production found by using the steps of 4.1.2.b with Step 3 replaced by:

"Step 3: If the home contains an indication-defining occurrence of the same terminal production of 'MODE mode indication' ('PRIORITY indication'), then the given occurrence identifies it; otherwise, Step 2 is taken."

{Indications have no inherent meaning. A terminal production of 'monadic indication' has no indication-defining occurrence.}

4.3. Operators

{Operators are either monadic-operators, i.e., require a right operand only, or are dyadic-operators, i.e., require both a left and a right operand, e.g., **abs** and / in **abs** x and x / y. Operators are made to possess routines by the elaboration of operation-declarations (7.5.1.a). Operators are identified by observing the modes of their operands, e.g., $x+y$, $x+i$, $i+x$, $i+j$ each involves a different operator, see 10.2.4.i, 10.2.5.a, 10.2.5.b and 10.2.3.i. Though the mode enveloped by the original of an operator contains the mode of the value, if any, delivered by its routine, this mode is not involved in the identification process.}

4.3.1. Syntax

a)⋆ operator : PRAM ADIC operator{b,c}.
b) procedure with LMODE parameter and RMODE parameter MOID PRIORITY operator{75b,84b} : PRIORITY indication{42e}.

c) procedure with RMODE parameter MOID monadic operator{75b,84g} : monadic indication{42f}.
d)★ dyadic operator : procedure with LMODE parameter and RMODE parameter MOID PRIORITY operator{b}.
e)★ monadic operator : procedure with RMODE parameter MOID monadic operator{c}.

{Examples:
b) + ;
c) **abs** }

4.3.2. Identifications of Operators

{The identification of operators is similar to that of identifiers and indications, except that different occurrences of one same terminal production of 'ADIC indication' may be occurrences of more than one terminal production of 'PRAM ADIC operator' and, therefore, the modes of the operands must be considered.}

a) A given occurrence of a terminal production of 'PRAM ADIC operator' where "PRAM" ("ADIC") stands for any terminal production of the metanotion 'PRAM' ('ADIC') is an operator-defining occurrence if it precedes the equals-symbol of an operation-declaration {7.5.1.a}; otherwise, it is an "operator-applied occurrence".

b) If a given {operator-applied} occurrence of a terminal production of 'PRAM ADIC operator' (see a) is the operator of a formula *F* {8.4.1.a}, then it may identify an operator-defining occurrence of the same terminal production found by using the steps of 4.1.2.b, with Step 3 replaced by:

"Step 3: If the home contains an operator-defining occurrence *O* {, in an operation-declaration (7.5.1.a,b),} of a terminal production *T* of 'PRAM ADIC operator' which is the same terminal production of 'ADIC indication' as the given occurrence, and which {, the identifications of all descendent identifiers, indications and operators of the operand(s) of *F* having been made,} is such that some formula exists which is the same sequence of symbols as *F*, whose operator is an occurrence of *T* and which is such that the original of each descendent identifier, indication and operator of its operand(s) is the same notion as the original of the corresponding identifier, indication and operator contained in *F* {, which, if the program is a proper program, is uniquely determined by virtue of 4.4.1.a}, then the given occurrence identifies *O*; otherwise, Step 2 is taken."

{Operators have no inherent meaning; an operator-defining occurrence is made to possess a routine (2.2.3.4) by the elaboration of an operation-declaration (7.5.1.a).

A given indication may be both a dyadic-indication and a dyadic-operator. As a dyadic-indication, it identifies its indication-defining occurrence. As a dyadic-operator, it may identify an operator-defining occurrence, which possesses a routine. Since the indication preceding the equals-symbol of an operation-declaration is an indication-application and an operator-definition (but not an operator-application), it follows that the set of those occurrences which identify a given dyadic-operator is a subset of those occurrences which identify the same dyadic-indication.

In the closed-clause

begin real *x, y* := *1.5;* **priority min** = *6;*
op min = *(***real** *a, b)* **real** : *(a>b | b | a); x* := *y* **min** *pi / 2* **end** ,

the first occurrence of **min** is an indication-defining priority-SIX-indication. The second occurrence of **min** is indication-applied and identifies the first occurrence (4.2.2), whereas, at the same textual position, **min** is also operator-defined as a [prrr]-priority-SIX-operator, where "[prrr]" stands for "procedure-with-real-parameter-and-real-parameter-real". The third occurrence of **min** is indication-applied and, as such, identifies the first occurrence, whereas, at the same textual position, **min** is also operator-applied, and, as such, identifies the second occurrence; this makes it, because of the identification condition (4.4.1.a), a [prrr]-priority-SIX-operator. This identification of the dyadic-operator is made because:

i) **min** occurs in an operation-declaration,

ii) the base *y* can be firmly coerced to the mode specified by **real**,

iii) the formula *pi / 2* is a priori of the mode specified by **real**,

iv) **min** is thus, because of the identification condition, a [prrr]-priority-SIX-operator.

If the first three conditions were not satisfied, then the search for an other defining occurrence would be continued in the same range, or failing that, in a surrounding range.}

4.4. Context Conditions

A "proper" program is a program satisfying the context conditions; a "meaningful" program is a proper program whose elaboration is defined by this Report. {Whether all programs, only proper programs, or only meaningful programs are "ALGOL 68" programs is a matter for individual taste. If one chooses only proper programs, then one may consider the context conditions as syntax which is not written as production rules.}

4.4.1. The Identification Conditions

a) In a proper program, a defining (indication-defining, operator-defining) occurrence of a terminal production of a notion ending with 'identifier' ('indication', 'operator') and each applied (indication-applied, operator-applied) occurrence identifying it are occurrences of one same terminal production of a notion ending with 'identifier' ('indication', 'operator').

b) No proper program contains an applied (indication-applied, operator-applied-occurrence of a terminal production of a notion ending with 'identifier' ('indica) tion', 'operator') which does not identify a defining (indication-defining, operator-defining) occurrence.

c) No proper program contains an indication which as an operator-applied occurrence identifies an operator-defining occurrence which as an indication-applied occurrence identifies an indication-defining occurrence different from the one identified by the given indication as an indication-applied occurrence.

{Condition c makes a program under certain circumstances improper independent of its elaboration. Without condition c, a program containing

(**priority** *? = 2;* **op** *? =* (**real** *a, b*) : **skip**;
(*random* < *0.5* | **priority** *? = 2; 0.1 ? 0.2*))

would be improper if, during the elaboration of this clause, the value of *random* < *0.5* turns out to be *true*. Then, the presence of an indication-defining occurrence of *?* in the serial-clause **priority** *? = 2; 0.1 ? 0.2* causes its protection (6.4.2.a, 6.1.2.a, 6.0.2.d) to replace both occurrences of *?* by another indication and thereby deprives the last occurrence of its operator-defining occurrence, which violates condition b. However, condition c makes the program improper immediately since the fourth occurrence of *?* identifies the third as its indication-defining occurrence and the second as its operator-defining occurrence which itself identifies the first occurrence as its indication-defining occurrence.}

4.4.2. The Uniqueness Conditions

a) A "reach" is a range {4.1.1.e} with the exclusion of all its constituent ranges.

b) No proper program contains a reach {a} containing two defining (indication-defining) occurrences of a given terminal production of a notion ending with 'identifier' ('indication').

{e.g., none of the closed-clauses (6.4.1.a)
(**real** *x;* **real** *x; sin (3.14)*),
(**real** *y;* **int** *y; sin (3.14)*),
(**real** *p; p:* **go to** *p; sin (3.14)*),
(**mode a** = **real**; **mode a** = **real**; *sin (3.14)*), or
(**priority b** = *5;* **priority b** = *6; sin (3.14)*)
is contained in a proper program.}

c) No proper program contains a reach containing two operation-declarations the operators of whose captions are the same terminal productions of a notion ending with 'indication' and all of whose correspondent constituent virtual-parameters {7.5.1.b, 7.1.1.x, 5.4.1.c, 7.1.1.y} are virtual-declarers specifying modes loosely related to one another {4.4.3.c}.

{e.g., neither the closed-clause
(**op max** = (**int** *a,* **int** *b*) **int** : (*a>b* | *a* | *b*);
op max = (**int** *a,* **int** *b*) **real** : (*a>b* | *a* | *b*); *sin (3.14)*)
nor
(**op max** = (**int** *a,* **ref int** *b*) **int** : (*a>b* | *a* | *b*);
op max = (**ref int** *a,* **int** *b*) **int** : (*a>b* | *a* | *b*); *sin (3.14)*)
is contained in any proper program, but
(**op max** = (**int** *a,* **int** *b*) **real** : (*a>b* | *a* | *b*);
op max = (**real** *a,* **real** *b*) **real** : (*a>b* | *a* | *b*); *sin (3.14)*)
may be.}

{In the pragmatic remarks in the sequel, "in the reach of (the declaration)" stands for "in a context where all identifications are made as in a reach containing (the declaration)".}

4.4.3. The Mode Conditions

a) A mode *M* is "strongly coerced from" ("firmly coerced from") a mode *N* if the notion '*N* base' is a production of the notion 'strong *M* base' ('firm *M* base') {see 8.2}; *M* is "united from" *N* if *M* is 'union of LMOODSETY *N* RMOODSETY mode' where "LMOODSETY" ("RMOODSETY") stands for any terminal production of the metanotion 'LMOODSETY' ('RMOODSETY').

{e.g., the mode specified by **real** is firmly coerced from the mode specified by **ref real** because the notion 'reference to real base' is a production of 'firm real base' (8.2.0.1.e, 8.2.1.1.a); the mode specified by **union** *(***int**, **real***)* is united from those specified by **int** and **real**.}

b) Two modes are "related" to one another if they are both firmly coerced {a} from one same mode. {A mode is related to itself.}

c) Two modes are "loosely related" if they either are related or are firmly coerced from 'row of LMODE' and 'row of RMODE' where "LMODE" and "RMODE" stand for different loosely related modes.

{e.g., the modes specified by **proc real** and **ref real** are related and, hence, loosely related and those specified by [ ] **real** and by [ ] **ref real** are loosely related but not related.}

d) No proper program contains a declarer {7.1.1.a} specifying a mode united from {a} two modes related {b} to one another, nor does it contain two declarers specifying modes united from two modes *P* and *Q* in which *P* and *Q* are in a different order.

{e.g., the declarer **union** *(***real**, **ref real***)* is not contained in any proper program, and **union** *(***int**, **real***)* and **union** *(***real**, **int***)* may be, but then specify the same mode; see the remarks at the end of 7.1.1.}

e) No proper program contains a declarer {7.1.1.a} the field-selectors {7.1.1.i} of two of whose constituent field-declarators {7.1.1.g} are the same sequence of symbols.

{e.g., the declarer **struct** *(***int** *i*, **bool** *i)* is not contained in any proper program, but **struct** *(***int** *i*, **struct** *(***int** *i*, **bool** *j) j)* may be.}

4.4.4. The Declaration Condition

a) A mode-indication {4.2.1.b} contained in a declarer {7.1.1.a} is "shielded by that declarer if

i) it is, or is contained in, a virtual-declarer {7.1.1.b} following a reference-to-symbol {3.1.1.d} in a field-declarator {7.1.1.g}, or

ii) it is, or is contained in, a virtual-declarer contained in a field-declarator contained in a virtual-declarer following a reference-to-symbol, or

iii) it is, or is contained in, a virtual-parameter {7.1.1.y}, or

iv) it is, or is contained in, a virtual-declarer following a virtual-parameters-pack {5.4.1.f}, or

v) it is, or identifies, an indication-defining occurrence contained in that declarer.

{e.g., **person** is shielded in **struct** *(***int** *age,* **ref person** *father),* but not in **struct** *(***int** *age,* **person** *uncle)* and **a** is shielded in **proc** *(***a***)* **a**, but not in **union** *(***int**, [ ] **a***).*}

b) An actual-declarer *D* {7.1.1.b} may "show" a mode-indication *M* {4.2.1.b}; this is determined in the following steps:

Step 1: *D* is protected and a copy is made of it; each mode-indication is said not to have been "encountered";

Step 2: If the copy is, or contains and does not shield {a}, a mode-indication which is the same terminal production as *M*, then *D* shows *M*; otherwise, Step 3 is taken;

Step 3: If the copy is, or contains and does not shield, a not yet encountered mode-indication *O*, then *O* and all mode-indications consisting of the same sequence of symbols are said to have been encountered, *O* is replaced by a copy of the protected actual-declarer of that mode-declaration whose mode-indication is the indication-defining occurrence identified by *O*, and Step 2 is taken; otherwise, *D* does not show *M*.

{e.g., in the declaration

mode a = [*1* : *2*] **b**, **b** = **union** *(***ref d**, **ref real***)*, **d** = **struct** *(***ref e** *e),*
e = **proc** *(***int***)* **a** ,

the mode-indications shown by [*1* : *2*] **b** are **b** and **d**.}

c) No proper program contains a mode-declaration {7.2.1.a} whose mode-indication is shown by its actual-declarer.

{e.g., none of the declarations

mode a = **a** ,
mode b = **e**, **e** = [*1* :*10*] **b** ,
mode d = [*1* : *2*] **ref union** *(***proc** *(***d***)* **d**, **proc d***)* ,
mode parson = **struct** *(***int** *age,* **parson** *uncle)*

is contained in any proper program.}

5. Denotations

{Denotations, e.g., *3.14* or *"abc"*, are terminal productions of notions whose value is independent of the elaboration of the program. In other languages, they are sometimes termed "literals" or "constants".}

5.0.1. Syntax

a)⋆ denotation : PLAIN denotation{510b,511a,512a,513a,514a} ;
BITS denotation{52b} ; row of character denotation{53b} ;
procedure with PARAMETERS MOID denotation{54b} ;
format denotation{55a}.

{Examples:

a) *3.14* ; **1 0 1** ; *"algol.report"* ; *((***bool** *a)* **int** : *(a | 1 | 0))* ; *$5d$*}

5.0.2. Semantics

Each denotation possesses a new instance of one same value whose mode is that enveloped {1.1.6.j} by its original {1.1.6.c}; its elaboration involves no action.

{e.g., the value of *"algol.report"* which is a production of 'row of character denotation' is of the mode 'row of character'.}

5.1. Plain Denotations

{Plain-denotations are those of arithmetic values, truth values and characters, e.g., *1*, *3.14*, **true** and *"a"*.}

5.1.0.1. Syntax

a)* plain denotation : PLAIN denotation{510b,511a,512a,513a,514a}.
b) long INTREAL denotation{860a} :
    long symbol{31d}, INTREAL denotation{511a,512a}.

{Examples:
b) **long** *0* ; **long long** *3.1415926535 8979323846 2643383279 5028841971 69399*}

5.1.0.2. Semantics

a) A plain-denotation possesses a plain value {2.2.3.1}, but plain values possessed by different plain-denotations are not necessarily different {e.g., *123.4* and *1.234e+2*}.

b) The value of a denotation consisting of a number {, possibly zero,} of long-symbols followed by an integral-denotation (real-denotation) is the "a priori" value of that integral-denotation (real-denotation) provided that it does not exceed the largest integer {10.1.b} (largest real number {10.1.d}) of length number one more than that number of long-symbols {; otherwise, the value is undefined}.

5.1.1. Integral Denotations

5.1.1.1. Syntax

a) integral denotation{510b,512c,d,f,i,55g,860a} : digit token{303c} sequence.

{Examples:
a) *0* ; *4096* ; *00123* (Note that *—1* is not an integral-denotation.)}

5.1.1.2. Semantics

The a priori value of an integral-denotation is the integer which in decimal notation is that integral-denotation in the representation language {1.1.8}. {See also 5.1.0.2.b.}

5.1.2. Real Denotations

5.1.2.1. Syntax

a) real denotation{510b,860a} :
    variable point numeral{b} ; floating point numeral{e}.
b) variable point numeral{a,f} : integral part{c} option, fractional part{d}.
c) integral part{b} : integral denotation{511a}.
d) fractional part{b} : point symbol{31b}, integral denotation{511a}.
e) floating point numeral{a} : stagnant part{f}, exponent part{g}.
f) stagnant part{e} : integral denotation{511a} ; variable point numeral{b}.

g) exponent part{e} : times ten to the power choice{h}, power of ten{i}.
h) times ten to the power choice{g} :
  times ten to the power symbol{31b} ; letter e{302b}.
i) power of ten{g} : plusminus{304c} option, integral denotation{511a}.

{Examples:
a) *0.000123* ; *1.23e−4* ;
b) *.123* ; *0.123* ;
c) *123* ;
d) *.123* ;
e) *1.23e−4* ;
f) *1* ; *1.23* ;
g) *e−4* ;
h) $_{10}$; *e* ;
i) *3* ; *+45* ; *−678* }

5.1.2.2. Semantics

a) The a priori value of a fractional-part is the a priori value of its integral-denotation divided by *10* as many times as there are digit-tokens in the fractional-part.

b) The a priori value of a variable-point-numeral is the sum in the sense of numerical analysis of *0*, the a priori value of its integral-part, if any, and that of its fractional-part {. See also 5.1.0.2.b}.

c) The a priori value of an exponent-part is *10* raised to the a priori value of the integral-denotation of its power-of-ten if that power-of-ten does not begin with a minus-symbol; otherwise, it is *1/10* raised to the a priori value of that integral-denotation.

d) The a priori value of a floating-point-numeral is the product in the sense of numerical analysis of the a priori values of its stagnant-part and exponent-part {. See also 5.1.0.2.b}.

5.1.3. Boolean Denotations

5.1.3.1. Syntax

a) boolean denotation{860a} : true symbol{31b} ; false symbol{31b}.

{Examples:
a) **true** ; **false** }

5.1.3.2. Semantics

The value of a true-symbol (false-symbol) is *true* (*false*).

5.1.4. Character Denotations

{Character-denotations consist of a string-item between two quote-symbols, e.g., *"a"*. To indicate a quote, a double quote-symbol is used for the string-item: *""""*. Since the syntax nowhere allows character- or string-denotations to follow one another, ambiguities do not arise.}

5.1.4.1. Syntax

a) character denotation{55j,k,860a} :
  quote symbol{31i}, string item{b}, quote symbol{31i}.

b) string item{a,53b} :
  character token{309d} ; quote image{c} ; other string item{1.1.5.c}.

c) quote image{b} : quote symbol{31i}, quote symbol{31i}.

{Examples:

a) *"a"* ;

b) *a* ; *""* ; *?* ;

c) *""*}

5.1.4.2. Semantics

a) Each string-item possesses a unique character. {The character possessed by a quote-image (space-symbol, digit-zero, digit-token, point-symbol, times-ten-to-the-power-choice, plus-i-times-symbol, plus-symbol) may be termed a quote (space, zero, digit, point, times ten to the power, plus i times, plus).}

b) The value of a character-denotation is a new instance of the character possessed by its string-item.

5.2. Bits Denotations

{There are two kinds of denotations of structured or multiple values, viz., bits-denotations, e.g., **1 0 1 1**, and string-denotations, e.g., *"abc"*. These denotations differ in that a string-denotation contains zero or two or more string-items but a bits-denotation may contain one or more flipflops. (See also character-denotations 5.1.4.)}

5.2.1. Syntax

a)★ bits denotation : BITS denotation{b,c}.

b) structured with row of boolean field LENGTH LENGTHETY letter aleph
  denotation{b,860a} : long symbol{31d}, structured with row of boolean field
  LENGTHETY letter aleph denotation{b,c}.

c) structured with row of boolean field letter aleph denotation{b,860a} :
  flipflop{303e} sequence.

{Examples:

b) **long 1 0 1 1** ;

c) **1 0 1 1**}

5.2.2. Semantics

Let *m* stand for the number of flipflops in the bits-denotation and *n* for the value of *L bits width* {10.1.g}, *L* standing for as many times *long* as there are long-symbols in the bits-denotation; if $m \leq n$, then the value of the bits-denotation is a structured value with one field selected by a field-selector which is the same sequence of symbols as *L* followed by letter-aleph, that field being a multiple value {2.2.3.3} whose descriptor has an offset *1* and one quintuple *(1, n, 1, 1, 1)* and whose element with index *j* is a new instance of *false* for $j = 1, \ldots, n - m$, and for $j = n - m + 1, \ldots, n$ is a new instance of *true* (*false*) if the *i*-th constituent flipflop ($i = j + m - n$) of the bits-denotation is a flip-symbol (flop-symbol).

5.3. String Denotations

5.3.1. Syntax

a)⋆ string denotation : row of character denotation{b}.

b) row of character denotation{55j,k,860a} : quote symbol{31i}, string item{514b} sequence proper option, quote symbol{31i}.

{Examples:

b) *""; "abc"; """"a.+.b"".is.a.formula"*}

5.3.2. Semantics

The value of a string-denotation is a multiple value {2.2.3.3} whose descriptor consists of an offset *1* and one quintuple *(1, n, 1, 1, 1)*, where *n* stands for the number of string-items contained in the string-denotation; for $i = 1, \ldots, n$, the element with index *i* of that multiple value is a new instance of the character possessed by the *i*-th constituent string-item of the string-denotation.

{The construction *"a"* is a character-denotation, not a string-denotation. However, in all strong positions, e.g., **string** *s* := *"a"*, it can be rowed to a multiple value (8.2.6). Elsewhere, where a multiple value is required, a cast (8.3.4.1.a) may be used, e.g., **union** *(***int**, **string***) is* := **string** : *"a"*. The "string", i.e., value of mode 'row of character', possessed by *""""a.+.b"".is.a.formula"* may well be presented informally as follows: "*a* + *b*" *is a formula*.}

5.4. Routine Denotations

{A routine-denotation, e.g., *((***real** *a, b)* **real** : *(a>b | b | a))*, always has a formal-parameters-pack, e.g., *(***real** *a, b)*. To the right of this formal-parameters-pack stands a cast, e.g., **real** : *(a>b | b | a)*, whose virtual-declarer specifies the mode of the value, if any, delivered by the elaboration of the routine, e.g., **real**. The whole is enclosed between an open-symbol and a close-symbol, but these may often be omitted, see the extension 9.2.d. It is essential that, in general, a routine-denotation be closed, for, otherwise, denotations like *(***int** *sintzoff)* : *(***int** *branquart)* : *lewi (wodon)* could also be calls, or formulas like *(***int** *a)* **int** : *1+2+3* would be ambiguous if + is also declared as an operator accepting a routine as left operand.}

5.4.1. Syntax

a)⋆ routine denotation : procedure with PARAMETERS MOID denotation{b}.

b) procedure with PARAMETERS MOID denotation{860a} : open symbol{31e}, formal PARAMETERS{c,e} pack, MOID cast{834a}, close symbol{31e}.

c) VICTAL PARAMETERS and PARAMETER{b,c,71x,862a} : VICTAL PARAMETERS{c,e,71y,74b}, gomma{d}, VICTAL PARAMETER{e,71y,74b}.

d) gomma{c} : go on symbol{31f} ; comma symbol{31e}.

e) formal MODE parameter{b,c,74a} : formal MODE declarer{71b}, MODE mode identifier{41b}.

f)⋆ VICTAL parameters pack : VICTAL PARAMETERS{c,e,71y,74b} pack.

{Examples:
b) *((***bool** *a, b)* **bool** : *(a | b |* **false***))* ;
c) *[1 :]* **real** *a; [1 :* ⌈ *a]* **real** *b* ;
d) *;; ,;*
e) **bool** *a* }

5.4.2. Semantics

A routine-denotation possesses that routine which can be obtained from it in the following steps:

Step 1: A copy is made of the routine-denotation;

Step 2: An equals-symbol followed by a skip-symbol is inserted in the copy following the last identifier in each copied constituent formal-parameter of the formal-parameters-pack of the routine-denotation; the open-symbol of that formal-parameters-pack is deleted and its close-symbol is replaced by a go-on-symbol;

Step 3: If the cast of the routine-denotation is a void-cast, then an open-symbol is inserted in the copy preceding, and a close-symbol following that cast; the copy, thus modified, is the routine possessed by the routine-denotation.

{The routine possessed by *p1* after the elaboration of the identity-declaration (7.4.1.a) **proc** *p1* = *(***int** *a, b)* **real** : *(a>b | xx | yy)* is *(***int** *a* = ∿, **int** *b* = ∿; **real** : *(a>b | xx | yy))* and that possessed by *p2* after the elaboration of **proc** *p2* = *(***real** *a;* **real** *b)* : *(a>b | stop)* is *(***real** *a* = ∿; **real** *b* = ∿; *(: (a>b | stop)))*. A routine is the same sequence of symbols as some closed-clause (6.3.1.a). For the use of routines, see 8.4 (formulas), 8.2.2 (deprocedured-coercends) and 8.6.2 (calls).}

5.5. Format Denotations

5.5.1. Syntax

a) format denotation{860a} :
  formatter symbol{31b}, collection{b} list, formatter symbol{31b}.
b) collection{a,b} : picture{c} ; insertion{d} option, replicator{f}, collection{b} list pack, insertion{d} option.
c) picture{b} : MODE pattern{552a,553a,554a,555a,556b,557b,—} option, insertion{d} option.
d) insertion{b,c,m,552b,f,554a,557b} :
  literal{j} option, insert{e} sequence ; literal{j}.
e) insert{d} : replicator{f}, alignment{i}, literal{j} option.
f) replicator{b,e,j,n} : replication{g} option.
g) replication{f,k,557b} :
  dynamic replication{h} ; integral denotation{511a}.
h) dynamic replication{g} :
  letter n{302b}, strong CLOSED integral clause{63a,640a,—}.
i) alignment{e} : letter k{302b} ; letter x{302b} ; letter y{302b} ; letter l{302b} ; letter p{302b}.

j) literal{d,e,552f,554b} : replicator{f}, STRING denotation{514a,53b}, replicated literal{k} sequence option.

k) replicated literal{j} : replication{g}, STRING denotation{514a,53b}.

{Examples:

a) *$p"table.of"x10a,n(lim−1)(16x3zd,3x3(2x+.12de+2d"+j×"si +.10de+2d)l)p$* ;

b) *p"table.of"x10a* ; *3x3(2x+.12de+2d"+j×"si+.10de+2d)l* ;

c) *120kc("mon","tues","wednes","thurs","fri","satur","sun")"day"* ; *p* ;

d) *p"table.of"x* ; *"day"* ;

e) *p"table.of"* ;

g) *n(lim−1)* ; *10* ;

h) *n(lim−1)* ;

j) *"+j×"* ;

k) *20"."* }

l) sign mould{552a,553a,d,e} : loose replicatable zero frame{m}, sign frame{p} ; loose sign frame{m}.

m) loose ANY frame{l,552d,553b,d,555a,556b,557b} : insertion{d} option, ANY frame{n,p,q,557c}.

n) replicatable ANY frame{m} : replicator{f}, ANY frame{o,q}.

o) zero frame{n,552e} : letter z{302b}.

p) sign frame{l,m} : plusminus{304c}.

q) suppressible ANY frame{m,n,557b} : letter s{302b} option, ANY frame{552e,553c,f,555b,556c}.

r)* frame : ANY frame{n,o,p,q,552e,553c,f,555b,556c,557c}.

{Examples:

l) *"="12z+* ; *2x+* ;

m) *"="12z* ;

n) *12z* ;

q) *si* ; *10a* }

{aa) Three ways of "transput" (i.e., "input" and "output") are provided by the standard-prelude, viz., formatless transput (10.5.2), formatted transput (10.5.3) and binary transput (10.5.4). Formats are used by the formatted-transput routines to control input from and output to a "file" (10.5.1). No section on semantics of format-denotations is given, since this is entirely dealt with by the standard-prelude.

bb) A format may be associated with a file by a call of *format* (10.5.3.a), *outf* (10.5.3.1.a) or *inf* (10.5.3.2.a), which causes a transformat to be elaborated (5.5.8.1.a), the collection-list of the format-denotation considered in 5.5.8.2.b.Step 2 to be unfolded (cc), the result to be the current picture-list of the file and its first constituent picture to be the current picture of the file (; e.g., after the call *format (f1, $pt,3(3d.d)l$)*, the current picture-list of the file *f1* is *pt, 3d.d, 3d.d, 3d.dl* and the current picture is *pt*).

cc) The result of unfolding a collection-list (10.5.3.b) is a picture-list obtained as follows:

a) if the collection-list is a picture, then the result consists of that picture;

b) if the collection-list is a collection but not a picture, then the result consists of the first insertion-option of the collection, followed by as many copies of the result of unfolding the collection-list of its collection-list-pack as is the value of its replicator, separated by comma-symbols, followed by its last insertion-option (; e.g., the result of unfolding *3k"ab"2(10a)l* is *3k"ab"10a, 10al*);

c) if the collection-list is a collection-list-proper, then the result consists of the result of unfolding the collection of that collection-list-proper followed by a comma-symbol, followed by the result of unfolding its collection-list (; e.g., the result of unfolding *10a,pn(i)(d.2d)"."* is *10a, p"."* when the value of *i* is *0*).

dd) When one of the formatted-transput routines *outf* (10.5.3.1.a), *out* (10.5.3.1.b), *inf* (10.5.3.2.a) or *in* (10.5.3.2.b) is called, then transput takes place in the following steps:

Step 1: The values to be transput are elaborated collaterally and the result is "straightened" (10.5.0.2) into a series of values, the first of which, if any, is made to be the current value;

Step 2: If the current picture of the file is an insertion-option, then its insertion, if any, is performed (gg), the next picture, if any, is made to be the current picture of the file and Step 2 is taken; otherwise, Step 3 is taken;

Step 3: If the series of values is empty or exhausted, then the transput is accomplished; otherwise, if the picture-list is exhausted, then *format end* of the file is called, a routine which may be provided by the programmer (10.5.1.kk);

Step 4: If the current value is "compatible" with (nn) the current picture, then that value is transput under control of that picture; otherwise, *value error* of the file is called, a routine which may be provided by the programmer;

Step 5: The next value, if any, is made to be the current value, the next picture, if any, is made to be the current picture and Step 2 is taken.

ee) The value of the empty replicator is *1*; the value of a replication which is an integral-denotation is the value of that denotation; the value of a dynamic-replication is the value of its integral-clause if that value is positive, and *0* otherwise.

ff) Transput occurs at the current "position" (i.e., page number, line number and char number) of the file. At each position of the file within certain limits (10.5.1.1.k,l,m) some character is "present", depending on the contents of the file and on its "conversion key" (10.5.1.ll).

gg) An insertion is performed by performing its constituent alignments and, on output (input), "writing" ("expecting") its constituent literals one after the other.

hh) Performing an alignment affects the position of the file as follows, where *n* stands for the value of the preceding replicator:

a) letter-k causes the current char number to be set to *n*;
b) letter-x causes the char number to be incremented by *n* (10.5.1.2.o);
c) letter-y causes the char number to be decremented by *n* (10.5.1.2.p);
d) letter-l causes the line number to be incremented by *n* and the char number to be reset to *1* (10.5.1.2.q);

e) letter-p causes the page number to be incremented by *n* and both the line number and the char number to be reset to *1* (10.5.1.2.r).

ii) A literal is written by writing the characters (strings) possessed by its constituent (row-of-)character-denotations each as many times as is the value of the preceding replicator; a string is written by writing its elements one after the other; a character is written by causing the character to be present at the current position of the file, thereby obliterating the character that was present, and then incrementing the char number by *1*. A literal is expected by expecting the characters (strings) possessed by its constituent (row-of-)character-denotations each as many times as is the value of the preceding replicator; a string is expected by expecting its elements one after the other; a character is expected by incrementing the char number by *1* if the character is present at the current position of the file; otherwise, the further elaboration is undefined.

jj) When a string whose number of characters is given is "read", then that number of characters are read and the result is a string whose elements are those characters; when a string is read under control of a given "terminator-string", then as long as the line is not exhausted, characters are read up to but not including the first character which is the same as some element of the terminator-string, and the result is a string whose elements are those characters; when a character is read, then the result is the character present at the current position of the file, and the char number of the file is incremented by *1*.

kk) The mode specified by a picture is that enveloped by the original of its pattern, if any. The number of characters specified by a picture is the sum of the numbers specified by its constituent frames and the number specified by a frame is equal to the value of its preceding replicator, if any, and *1* otherwise.

ll) On output, a picture may be used to "edit" a value in the following steps:

Step 1: The value is converted by an appropriate output routine (10.5.2.1.c,d,e) to a string of as many characters as specified by the picture (; if the pattern of the picture is an integral-pattern, then this conversion takes place to a base equal to the value of the integral-denotation which is the same sequence of symbols as its constituent radix, if any, and base *10* otherwise); if this number of characters is not sufficient, then *value error* of the file is called, a routine which may be provided by the programmer (10.5.1.kk);

Step 2: In those parts, if any, of the string specified by a sign-mould, a character specified by the sign-frame will be used to indicate the sign, viz., if the sign-frame is a minus-symbol and the value is nonnegative, then a space, and, otherwise, the character specified by the sign-frame; this character is shifted in that part of the string specified by the sign-mould as far to the right as possible across all leading zeroes, and those zeroes are replaced by spaces (; e.g., under the sign-mould *4z+*, the string possessed by *"+0003"* becomes that possessed by *"...+3"*); if the picture does not contain a sign-mould and the value is negative, then *value error* of the file is called;

Step 3: Leading zeroes in those parts of the string specified by any remaining zero-frames are replaced by spaces (; e.g., under the picture *zdzd2d*, the integer *180168* becomes the string possessed by *"18.168"*;

Step 4: For all frames occurring in the picture, first the preceding insertion, if any, is performed, and next, if the frame is not "suppressed" (, i.e., preceded by letter-s), then that part of the string specified by the frame is written; finally, the insertion, if any, following the last constituent frame is performed (; e.g., editing under the picture *zd"—"zd"—19"2d*, the integer ***180168*** causes the string possessed by *"18—.1—1968"* to be written).

mm) On input, a picture may be used to "indit" a value of a given mode from a file in the following steps:

Step 1: A string is obtained consisting of the characters obtained by performing the following process for all frames occurring in the picture, viz., first, the insertion, if any, preceding the frame is performed and next, as many characters are obtained as are specified by the frame; each of those characters is obtained,

if the frame is not suppressed, then by reading from the file a character, and, if the frame is a digit- (point-, exponent-, complex-)frame and the character is not a digit (point, times ten to the power, plus i times), then calling *char error* of the file (10.5.1.kk) with as its parameter a zero (point, times ten to the power, plus i times), and

if the frame is suppressed, then by taking, if the frame is a digit- (zero-, point-, exponent-, complex-, character-)frame a zero (zero, point, times ten to the power, plus i times, space);

Step 2: Those parts, if any, of the string specified by a sign-mould must contain a character, specified by the sign-frame, to indicate the sign (; see ll.Step 2); if those parts contain such a character, with only spaces appearing in front of it and no leading zeroes appearing after it, then those leading spaces, if any, are deleted; otherwise, *char error* is called with a plus; if this character is a space, and the sign-frame is a minus-symbol, then it is replaced by a plus (; e.g., if in Step 1 under control of *3z—d*, the string possessed by *"...39"* is obtained, then in Step 2 that possessed by *"+39"* is obtained);

Step 3: Leading spaces in those parts of the string specified by any remaining zero-frames are replaced by zeroes;

Step 4: The string is converted by an appropriate input routine (10.5.2.2.c,d,e) into a value of the given mode, if possible, and, otherwise, *value error* of the file is called (; e.g., if the value of *maxint* (10.1.b) is ***10000***, then under *+5d* it is possible to input +***10000***, but not +***10001***).

nn) A value of a given mode is compatible with a given picture if

a) on output, there exists some mode which is the mode specified by the picture preceded by zero or more times 'long', such that that mode is strongly coerced from the given mode;

b) on input, there exists some mode which is the mode specified by the picture preceded by 'reference to' followed by zero or more times 'long', such that that mode is strongly coerced from the given mode. (A value of mode 'reference to long integral' is on output compatible with a picture that specifies the mode 'real', but not on input.)

oo) Formats have a complementary meaning on input and output, i.e., a given value which is not a string with one or two flexible bounds, which has been output successfully to the file, under control of a certain picture, starting from a certain position, can be successfully input again from that file under control of the same picture, starting at the same position, provided that the contents of the file are not changed in between; if the picture does not contain a letter-k or letter-y as alignment, and the picture does not contain any digit-frames or character-frames preceded by letter-s, then the second value, obtained on input, is equal (approximately equal) to the given value if this is a string, integer or truth value (is a real number); output of this second value to the file has the same effect on the contents of the file as output of the given value under control of the same given picture and starting from one same position.

pp) If a value is transput under control of a picture whose constituent pattern is not an integral-choice-pattern (5.5.2.f), boolean-pattern (5.5.4.a) or string-pattern (5.5.7.b), then on output (input) it is edited (indited) under control of the picture.}

5.5.2. Syntax of Integral Patterns

a) integral pattern{55c} : radix mould{b} option, sign mould{55l} option, integral mould{d} ; integral choice pattern{f}.
b) radix mould{a} : insertion{55d} option, radix{c}, letter r{302b}.
c) radix{b} : digit two{303d} ; digit four{303d} ; digit eight{303d} ; digit one{303d}, digit zero{303d} ; digit one{303d}, digit six{303d}.
d) integral mould{a,553b,d,e} : loose replicatable suppressible digit frame{55m} sequence.
e) digit frame{55q} : zero frame{55o} ; letter d{302b}.
f) integral choice pattern{a} : insertion{55d} option, letter c{302b}, literal{55j} list pack.

{Examples:
a) *2r6d30sd* ; *12z+d* ; *zd"—"zd"—19"2d* ; *120kc("mon","tues","wednes","thurs","fri","satur","sun")* ;
b) *2r* ;
c) *2* ; *4* ; *8* ; *10* ; *16* ;
d) *zd"—"zd"—19"2d* ;
f) *120kc("mon","tues","wednes","thurs","fri","satur","sun")* }

{If a given value is transput under control of a picture whose constituent pattern is an integral-choice-pattern, then the insertion, if any, preceding the letter-c is performed, and,
a) on output, letting ***n*** stand for the integer to be output, if ***n*** > ***0*** and the number of literals in the constituent literal-list-pack is at least ***n***, then the ***n***-th literal is written on the file; otherwise, the further elaboration is undefined;
b) on input, one of the constituent literals of the constituent literal-list-pack is expected on the file; if the *i*-th constituent literal is the first one present, then the value is *i*; if none of these literals is present, then the further elaboration is undefined;
c) finally, the insertion, if any, following the pattern is performed.}

5.5.3. Syntax of Real Patterns

a) real pattern{55c,556b} : sign mould{551} option, real mould{b} ;
floating point mould{d}.

b) real mould{a,e} : integral mould{552d},
loose suppressible point frame{55m}, integral mould{552d} option ;
loose suppressible point frame{55m}, integral mould{552d}.

c) point frame{55q} : point symbol{31b}.

d) floating point mould{a} : stagnant mould{e}, loose suppressible exponent
frame{55m}, sign mould{551} option, integral mould{552d}.

e) stagnant mould{d} : sign mould{551} option, INTREAL mould{b,552d,—}.

f) exponent frame{55q} : letter e{302b}.

{Examples:

a) *+.12d* ; *+d.11de+2d* ;

b) *d.11d* ; *.12d* ;

d) *+d.11de+2d* ;

e) *+d.11d* }

5.5.4. Syntax of Boolean Patterns

a) boolean pattern{55c} : insertion{55d} option, letter b{302b},
boolean choice mould{b} option.

b) boolean choice mould{a} : open symbol{31e}, literal{55j},
comma symbol{31e}, literal{55j}, close symbol{31e}.

{Examples:

a) *l"result"14xb* ; *b("","error")* ;

b) *("","error")* }

{If the boolean-pattern does not contain a choice-mould, then the effect of using the pattern is the same as if the letter-b were followed by *("**1**","**0**")*. If a given value is transput under control of a picture whose constituent pattern is a boolean-pattern, then the insertion, if any, preceding the letter-b is performed, and,

a) on output, if the truth value to be output is *true*, then the first constituent literal of the constituent choice-mould is written, and, otherwise, the second;

b) on input, one of the constituent literals of the constituent choice-mould is expected on the file; if the first literal is present, then the value *true* is found; otherwise, if the second literal is present, then the value *false* is found; otherwise, the further elaboration is undefined;

c) finally, the insertion, if any, following the pattern is performed.}

5.5.5. Syntax of Character Patterns

a) character pattern{55c} : loose suppressible character frame{55m}.

b) character frame{55q} : letter a{302b}.

{Example:

a) *"."a* }

5.5.6. Syntax of Complex Patterns

a)* complex pattern : COMPLEX pattern{b}.

b) COMPLEX pattern{55c} : real pattern{553a},
loose suppressible complex frame{55m}, real pattern{553a}.

c) complex frame{55q} : letter i{302b}.

{Example:

b) *2x+.12de+2d"+j×"si+.10de+2d* }

5.5.7. Syntax of String Patterns

a)* string pattern : row of character pattern{b}.

b) row of character pattern{55c} : loose string frame{55m} ;
loose replicatable suppressible character frame{55m} sequence proper ;
insertion{55d} option, replication{55g}, suppressible character frame{55q}.

c) string frame{55m} : letter t{302b}.

{Examples:

b) *lt* ; *5a3sa5a* ; *p"table.of"x10a* (Note that *a* is a character-pattern, whereas *1a* is a string-pattern for a string with one element.) }

{If a given value is transput under control of a picture whose constituent pattern is a string-pattern, then, if the pattern is a loose-string-frame, then

a) the constituent insertion, if any, is performed;

b) on output, the given string is written on the file;

c) on input, if the string has fixed bounds, then that number of characters are read; otherwise, a string is read under control of the terminator-string of the file (10.5.1.mm);

d) finally, the insertion, if any, following the pattern is performed; otherwise, on output, the given string, which must have as many elements as the number of characters specified by the picture, is edited;
on input, the string is indited.}

5.5.8. Transformats

{Transformats are exclusively used as actual-parameters of formatted-transput routines; for reasons of efficiency, the programmer has deliberately been made unable to use them elsewhere by the choice of the field-selector, which contains letter-aleph for which no representation is provided. Although transformats are not denotations at all, they are handled here because of their close connection to formats.}

5.5.8.1. Syntax

a) structured with row of character field letter aleph digit one
transformat{74b} : firm format unit{61e}.

{Example:

a) *(x ≥ 0 | $5d$ | $5d"—"$)* }

5.5.8.2. Semantics

a) The format {2.2.3.4} possessed by a given format-denotation is the same sequence of symbols as the given format-denotation.

b) A given transformat is elaborated in the following steps:

Step 1: It is preelaborated {1.1.6.i};

Step 2: A format-denotation is considered which is the same sequence of symbols as the format obtained in Step 1;

Step 3: All constituent dynamic-replications {5.5.1.h} of the considered format-denotation are elaborated collaterally {6.2.2.a}, where the elaboration of a dynamic-replication is that of its integral-clause;

Step 4: Each of those dynamic-replications is replaced by an integral-denotation {5.1.1.1.a} which possesses the same value as that dynamic-replication if that value is positive, and, otherwise, by a digit-zero; furthermore, every replicator which is empty is replaced by a digit-one;

Step 5: That string-denotation {5.3.1.a} (character-denotation {5.1.4.1.a}) is considered which is obtained by replacing in the considered format-denotation as modified in Step 4 each constituent quote-symbol by a quote-image {5.1.4.1.c} and the first and the last constituent formatter-symbol by a quote-symbol;

Step 6: A new instance of the value of the considered string-denotation (of a multiple value composed of the value of the considered character-denotation as its {only} element and of a descriptor consisting of an offset *1* and one quintuple *(1, 1, 1, 1, 1)*) is made to be the {only} field of a new instance of a structured value {2.2.3.2} whose mode is that enveloped {1.1.6.j} by the original {1.1.6.c} of the transformat;

Step 7: The transformat is made to possess the structured value obtained in Step 6.

6. Phrases

{A phrase is a declaration or a clause. Declarations may be unitary, e.g., **real** *x*, or collateral, e.g., **real** *x, y*. Clauses may be unitary, e.g., *x* := *1*, collateral, e.g., *(x := 1, y := 2)*, closed, e.g., *(x + y)*, or conditional, e.g., **if** *x* > *0* **then** *x* **else** *0* **fi** (which may be written *(x > 0 | x | 0)*). Most clauses will be of a certain "sort", i.e., strong, weak, firm or soft, which determines how the coercions should be effected. The sort is "passed on" in the production rules for clauses and may be modified by "balancing" in serial-, collateral- and conditional-clauses.}

6.0.1. Syntax

a)* SOME phrase:
  SORTETY SOME PHRASE{61a,62a,b,c,d,f,63a,64a,c,d,e,70a,81a,—}.

b)* SOME expression:
  SORTETY SOME MODE clause{61a,62b,c,d,f,63a,64a,c,d,e,81a}.

c)* SOME statement: strong SOME void clause{61a,62b,63a,64a,c,e,81a}.

{The rules b and c are not actually used in this Report but serve to help the reader, who may know some such constructions in other languages under those appellations. For an informal introduction into ALGOL 68 (0.1.1) also the following rules may be helpful:

d)⋆ MODE constant: MODE FORM{830a,84b,g,850a,860a}.
e)⋆ MODE variable: reference to MODE FORM{830a,84b,g,850a,860a}.
f)⋆ procedure: REFETY PROCEDURE FORM{830a,84b,g,850a,860a}.
g)⋆ structure display:
strong collateral structured with FIELDS and FIELD clause{62f}.
h)⋆ row display: SORTETY collateral row of MODE clause{62c,d,—}.}

6.0.2. Semantics

a) The elaboration of a phrase begins when it is initiated, it may be "interrupted", "halted" or "resumed", and it ends by being "terminated" or "completed", whereupon, if the phrase "appoints" a unitary-phrase as its "successor", then the elaboration of that unitary-phrase is initiated.

b) The elaboration of a phrase may be interrupted by an action {e.g., "overflow"} not specified by the phrase but taken by the computer if its limitations {2.3.b} do not permit satisfactory elaboration. {Whether, after an interruption, the elaboration of the phrase is resumed, the elaboration of some unitary-phrase is initiated or the elaboration of the program ends, is left undefined in this Report.}

c) The elaboration of a phrase may be halted {10.4.c}, i.e., no further actions constituting the elaboration of that phrase take place until the elaboration of the phrase is resumed {10.4.d}, if at all.

d) A given {serial-}clause is "protected" in the following steps:

Step 1: If the given clause contains a defining occurrence {4.1.2.a} (an indication-defining occurrence {4.2.2.a}) of a terminal production of a notion ending with 'identifier' ('indication') which also occurs outside it, then that defining (indication-defining) occurrence and all occurrences identifying it are replaced by occurrences of one same terminal production of that notion which does not occur in the program and Step 1 is taken; otherwise, Step 2 is taken;

Step 2: If the given clause as possibly modified in Step 1, 2 or 4 contains an operator-defining occurrence {4.3.2.a} of a terminal production of a notion ending with 'indication' which also occurs outside it, then that operator-defining occurrence and all occurrences identifying it are replaced by occurrences of one same new terminal production of that notion which does not occur in the program and Step 3 is taken; otherwise, the protection of the given clause is accomplished;

Step 3: If the indication is a dyadic-indication, then Step 4 is taken; otherwise, Step 2 is taken;

Step 4: A copy is made of the priority-declaration containing the indication which, before the replacement in Step 2, was identified by that operator-defining occurrence; that indication in the copy is replaced by an occurrence of the new terminal production; the given clause is modified by inserting before it the thus modified copy of the priority-declaration followed by a go-on-symbol, and Step 2 is taken.

{Clauses are protected in order to allow unhampered definitions of identifiers, indications and operators within ranges and to permit a meaningful call, within a range, of a procedure declared outside it.}

{What's in a name? that which we call a rose
By any other name would smell as sweet.
Romeo and Juliet, William Shakespeare.}

6.1. Serial Clauses

{Serial-clauses are built from unitary-clauses and declarations with the help of go-on-symbols (;), completion-symbols (. or **exit**) and labels, e.g., *(x>0 | x:= 1 | l); y. l: y+1*, where the value of the clause is that of *y*, if *x>0* and that of *y+1* otherwise. A serial-clause may begin with declaration-preludes, e.g., **int** *n:= 1;* in **int** *n:= 1; x:= y+n.* Labels may occur in only three syntactic positions within serial-clauses: after a completion-symbol (here a label is obligatory, e.g., *.l:*), in a sequencer (e.g., *;l:*), or at the beginning of a clause-train (i.e., one or more unitary-clauses separated by sequencers, e.g., *l: x:= 1; y:= 1*). A declaration-prelude may begin with void-clauses (statements), e.g., in order to supply a multiple value as in *[1:n]* **real** *x1;* **for** *i* **to** *n* **do** *x1 [i] := i×i;* **real** *y;* ; however, these void-clauses may not be labelled. A declaration-prelude always ends with a go-on-symbol. The modes of some serial-clauses must be balanced (6.1.1.g). For remarks concerning the balancing of modes see 6.4.1.}

6.1.1. Syntax

a) SORTETY serial CLAUSE{63a,64b,e} : declaration prelude{b} sequence option, suite of SORTETY CLAUSE trains{f,g}.
b) declaration prelude{a,2b,c} :
 statement prelude{c} option, single declaration{d}, go on symbol{31f}.
c) statement prelude{b} : chain of strong void units{e} separated by go on symbols{31f}, go on symbol{31f}.
d) single declaration{b} :
 unitary declaration{70a} ; collateral declaration{62a}.
e) SORTETY MOID unit{c,h,i,558a,62b,c,e,h,74b,831c,834a} :
 SORTETY unitary MOID clause{81a}.
f) suite of STRONGETY CLAUSE trains{a,g} :
 chain of STRONGETY CLAUSE trains{h} separated by completers{l}.
g) suite of FEAT CLAUSE trains{a,g} : FEAT CLAUSE train{h} ;
 FEAT CLAUSE train{h}, completer{l}, suite of strong CLAUSE trains{f} ;
 strong CLAUSE train{h}, completer{l}, suite of FEAT CLAUSE trains{g}.
h) SORTETY MOID clause train{f,g,2g} : label{k} sequence option, statement interlude{i} option, SORTETY MOID unit{e}.
i) statement interlude{h,2f} :
 chain of strong void units{e} separated by sequencers{j}, sequencer{j}.
j) sequencer{i,30c} : go on symbol{31f}, label{k} sequence option.
k) label{h,j,l,2d} : label identifier{41b}, label symbol{31e}.
l) completer{f,g,30c} : completion symbol{31f}, label{k}.

{Examples:

a) **real** *a* := *0; l1: l2: x* := *a* + *1; (p | l3); (x>0 | l3 | x* := *1 − x);*
false. *l3: y* := *y* + *1;* **true**;

b) **real** *a* := *0; ; read (n); [1:n]* **real** *x1, y1; ;*

c) *read (n); ;*

d) **real** *a* := *0; [1:n]* **real** *x1, y1;*

e) **false**;

f) *l1: l2: x* := *a* + *1; (p | l3); (x>0 | l3 | x* := *1 − x);* **false**.
l3: y := *y* + *1;* **true**;

h) *l1: l2: x* := *a* + *1; (p | l3); (x>0 | l3 | x* := *1 − x);* **false**;

i) *x* := *a* + *1; (p | l3); (x>0 | l3 | x* := *1 − x); ;*

j) *; ; ; l4: l5: ;*

k) *l4: ;*

l) *. l3:* }

6.1.2. Semantics

a) The elaboration of a serial-clause is initiated by protecting {6.0.2.d} it and then initiating the elaboration of its textually first constituent single-declaration or unitary-clause.

b) The completion of the elaboration of a single-declaration or unitary-clause preceding a go-on-symbol followed (not followed) by a label-sequence initiates the elaboration of the unitary-clause following that label-sequence (the single-declaration or unitary-clause following that go-on-symbol).

c) The elaboration of a serial-clause is

interrupted (halted, resumed) upon the interruption (halting, resumption) of the elaboration of a constituent single-declaration or unitary-clause;

terminated upon the termination of the elaboration of a constituent single-declaration or unitary-clause appointing a successor outside the serial-clause, and that successor {8.2.7.2.b.Step 1} is appointed the successor of the serial-clause.

d) The elaboration of a serial-clause is completed upon the completion of the elaboration of its textually last constituent unitary-clause or of that of a constituent unitary-clause preceding a completer.

e) The value of a serial-clause is the value of that constituent unitary-clause the completion of whose elaboration completed the elaboration of the serial-clause provided that the scope {2.2.4.2} of that value is larger than the serial-clause {; otherwise, the value of the serial-clause is undefined}.

{In *y* := *(x* := *1.2; 3.4)* the value of the serial-clause *x* := *1.2; 3.4* is the real number possessed by *3.4*. In *xx* := *(***real** *r* := *0.1; r)*, the value of the serial-clause **real** *r* := *0.1; r* is undefined since the scope of the name possessed by *r* is the serial-clause itself, whereas, in *y* := *(***real** *r* := *0.1; r)*, the serial-clause **real** *r* := *0.1; r* possesses a real number.}

6.2. Collateral Phrases

{Collateral-phrases contain two or more unitary-phrases separated by comma-symbols (,) and, in the case of collateral-clauses, are enclosed between an open-symbol (*(*) and a close-symbol (*)*) or between a begin-symbol (**begin**) and an end-symbol (**end**), e.g., *(x := 1, y := 2)* or **real** *x*, **real** *y* (usually **real** *x, y*, see 9.2.c). The values of collateral-clauses which are not statements (void-clauses) are either multiple or structured values, e.g., *(1.2, 3.4)* in [ ] **real** *x1* = *(1.2, 3.4)* and in **compl** *z* := *(1.2, 3.4)*. Here, the collateral-clause *(1.2, 3.4)* obtains the mode 'row of real' or the mode which is the terminal production of 'COMPLEX'. Collateral-clauses whose value is structured must contain at least two fields, for, otherwise, in the reach of the declarations **struct m** = *(***ref m** *m)*; **m** *nobuo*, *yoneda*, the assignation *nobuo* := *(yoneda)* would be syntactically ambiguous. In the reach of the declarations **struct r** = *(***real** *a)*; **r** *r*, the construction *r* := *(3.14)* is not an assignation, but *a* **of** *r* := *3.14* is. It is possible to present a single value or no value at all as a multiple value, e.g., [ ] **real** *x1* = ; *[1 : 1]* **real** *y1* := *3*, but this involves a coercion known as rowing; see 8.2.6.}

6.2.1. Syntax

a) collateral declaration{61d} : unitary declaration{70a} list proper.

b) strong collateral void clause{2d,81d} :
   parallel symbol{31e} option, strong void unit{61e} list proper PACK.

c) STRONGETY collateral row of MODE clause{81d} :
   STRONGETY MODE unit{61e} list proper PACK.

d) firm collateral row of MODE clause{81d} : firm MODE balance{e} PACK.

e) firm MODE balance{d,e} :
   firm MODE unit{61e}, comma symbol{31e}, strong MODE unit{61e} list ;
   strong MODE unit{61e}, comma symbol{31e}, firm MODE unit{61e} ;
   strong MODE unit{61e}, comma symbol{31e}, firm MODE balance{e}.

f) strong collateral structured with FIELDS and FIELD clause{81d} :
   strong structured with FIELDS and FIELD structure{g} PACK.

g) strong structured with FIELDS and FIELD structure{f,g} :
   strong structured with FIELDS structure{g,h}, comma symbol{31e},
   strong structured with FIELD structure{h}.

h) strong structured with MODE field TAG structure{g} :
   strong MODE unit{61e}.

{Examples:

a) **real** *x*, **real** *y* ; (and by 9.2.c) **real** *x, y* ;

b) *(x := 1, y := 2, z := 3)* ;

c) *(x, n)* ;

d) *(1.2, 3, 4)* (in *(1.2, 3, 4)* + *x1*, supposing + has been declared also for 'row of real') ;

e) *1.2, 3, 4* (in *(1.2, 3, 4)* + *x1*) ; *1, 2.3* (in *(1, 2.3)* + *x1*) ;
   1, 2.3, 4 (in *(1, 2.3, 4)* + *x1*) ;

f) *(1, 2.3)* (in *z* := *(1, 2.3)*) ;

g) *1, 2.3* ;

h) *1* }

6.2.2. Semantics

a) If constituents of an occurrence of a terminal production of a notion are "elaborated collaterally", then this elaboration is the collateral action {2.2.5} consisting of the {merged} elaborations of these constituents, and is

initiated by initiating the elaboration of each of these constituents,

interrupted upon the interruption of the elaboration of any of these constituents,

completed upon the completion of the elaboration of all of these constituents, and

terminated upon the termination of the elaboration of any of these constituents, and if that constituent appoints a successor, then this is the successor of the occurrence.

b) A collateral-declaration is elaborated by elaborating its constituent unitary-declarations collaterally {a}.

c) A collateral-clause is elaborated in the following steps:

Step 1: Its constituent units are elaborated collaterally {a};

Step 2: If the terminal production of the metanotion 'MOID' enveloped {1.1.6.j} by the original {1.1.6.c} of the collateral-clause is a mode, then this mode is considered and Step 3 is taken; otherwise, {it is 'void' and} the elaboration of the collateral-clause is complete;

Step 3: If one of the values of the units obtained in Step 1 is a name {2.2.3.5} which refers to a component of a multiple value having one or more states {2.2.3.3} equal to *0*, then the further elaboration is undefined; otherwise, Step 4 is taken;

Step 4: If the considered mode begins with 'row of', then Step 5 is taken; otherwise, new instances of the values obtained in Step 1 are made, in the given order, to be the fields of a new instance of a structured value {2.2.3.2}; this structured value is considered and Step 7 is taken;

Step 5: Let m stand for the number of constituent units in the collateral-clause; if the considered mode begins with 'row of row of', then Step 6 is taken; otherwise, a new instance of a multiple value is created whose element with index i is a new instance of the value of the i-th constituent unit, and whose descriptor consists of an offset *1* and one quintuple $(1, m, 1, 1, 1)$; this multiple value is considered and Step 7 is taken;

Step 6: If not all corresponding upper (lower) bounds of the multiple values obtained in Step 1 are equal, then the further elaboration is undefined; otherwise, the elements with indices $(i-1)\times r+j$, $j=1, \ldots, r$ of the new value, where r stands for the number of elements in one of those values, are new instances of the elements, taken in the order of ascending indices, of the value of the i-th constituent unit and the descriptor of the new value is a copy of the descriptor of the value of one of the constituent units into which an additional quintuple $(1, m, 1, 1, 1)$ has been inserted before the old first quintuple, the offset has been set to *1*, d_n has been set to *1*, and for $i=n, n-1, \ldots, 2$, the stride d_{i-1} has been set to $(u_i-l_i+1)\times d_i$; this new multiple value is considered and Step 7 is taken;

Step 7: The value of the collateral-clause is the considered value; its mode is the considered mode.

6.3. Closed Clauses

{Closed-clauses are generally used to construct primaries (8.1.1.d) from serial-clauses, e.g., *(x+y)* in *(x+y)×a*. The question of identification (Chapter 4) and protection (6.0.2.d) may arise in closed-clauses, because a serial-clause is a range (4.1.1.e) and it may begin with a declaration-prelude (6.1.1.a).}

6.3.1. Syntax

a) SORTETY closed CLAUSE{2d,55h,81d} : SORTETY serial CLAUSE{61a} PACK.

{Examples:

a) **begin** *i:=i+1; j:=j+1* **end**; *(x+y)* }

6.3.2. Semantics

The elaboration of a closed-clause is that of its constituent serial-clause, and its value is that, if any, of that serial-clause.

6.4. Conditional Clauses

{Conditional-clauses allow the programmer to choose one out of a pair of clauses, depending on the value (which is of mode 'boolean') of a condition, e.g., *(x>0 | x | 0)*. Here, *x>0* is the condition. If its value is *true*, then *x*, and, otherwise, *0* is chosen. Conditional-clauses are generalized in the extensions 9.4, e.g., **if** *x>0* **then** *x* **elsf** *x<−1* **then** *−x−1* **else** *0* **fi**, which has the same effect as *(x>0 | x | (x<−1 | −x−1 | 0))*. Unlike similar constructions in other languages, conditional-clauses are always enclosed between an if-symbol, represented by *(*, by **if** or by **case**, and a fi-symbol, represented by *)*, by **fi** or by **esac**. This enclosure allows both parts of the choice-clause and the condition to contain serial-clauses.}

6.4.1. Syntax

a) SORTETY conditional CLAUSE{2d,55h,81d} : if symbol{31e}, condition{b}, SORTETY choice CLAUSE{c,d}, fi symbol{31e}.

b) condition{a} : strong serial boolean clause{61a}.

c) STRONGETY choice CLAUSE{a} :
STRONGETY then CLAUSE{e}, STRONGETY else CLAUSE{e}.

d) FEAT choice CLAUSE{a} :
FEAT then CLAUSE{e}, strong else CLAUSE{e} ;
strong then CLAUSE{e}, FEAT else CLAUSE{e}.

e) SORTETY THELSE CLAUSE{c,d} :
THELSE symbol{31e}, SORTETY serial CLAUSE{61a}.

{Examples:

a) *(x>0 | x | 0)* ; **if** *overflow* **then** *exit* **fi** (see 9.4.a) ;

b) *x>0* ; *overflow* ;

c) *| x | 0* ; **then** *exit* ;

d) | *x* | *0* (in *(x>0* | *x* | *0)* +*y*) ;
e) | *x* ; | *0* ; **then** *exit* }

{Rule d illustrates why modes should be balanced (see also 6.1.1.g and 6.2.1.e). Thus, if a choice-clause is, say, firm, then at least one of its two constituent clauses must be firm, while the other may be strong. For example, in *(p* | *x* | ∿*)* + *(p* | ∿ | *y)*, the conditional-clause *(p* | *x* | ∿*)* is balanced by making | *x* firm and | ∿ strong, whereas *(p* | ∿ | *y)* is balanced by making | ∿ strong and | *y* firm. The example *(p* | ∿ | ∿*)* +*y* illustrates that not both may be strong, for otherwise the operator + could not be identified.}

6.4.2. Semantics

a) A conditional-clause is elaborated in the following steps:

Step 1: Its condition is elaborated;

Step 2: If the value of that condition is ***true*** (***false***) then the serial-clause of the then-clause (else-clause) of its choice-clause is considered;

Step 3: The considered clause is elaborated;

Step 4: The value, if any, of the conditional-clause, then is that of the clause elaborated in Step 3.

b) The elaboration of a conditional-clause is

interrupted (halted, resumed) upon the interruption (halting, resumption) of the elaboration of its condition or the considered clause;

completed upon the completion of the elaboration of the considered clause;

terminated upon the termination of the elaboration of its condition or the considered clause, and if one of these appoints a successor, then this is the successor of the conditional-clause.

7. Unitary Declarations

{Unitary-declarations provide the indication-defining occurrences of mode-indications, e.g., **string** in **mode string** = [*1* : *0* **flex**] **char**, and of dyadic-indications, e.g., ∨ in **priority** ∨ =*2*, the defining occurrences of mode-identifiers, e.g., *x* in **real** *x*, and the operator-defining occurrences of operators, e.g., **abs** in **op abs** = *(***int** *a)* **int** : *(a<0* | −*a* | *a)*. Declarations occur in declaration-preludes (6.1.1.b).}

7.0.1. Syntax

a) unitary declaration{61d,62a} :
mode declaration{72a} ; priority declaration{73a} ;
identity declaration{74a} ; operation declaration{75a}.

{Examples:

a) **mode string** = [*1* : *0* **flex**] **char** ; **priority** ∨ =*2* ;
int *m* =*4096* ; **op** ÷ = *(***real** *a, b)* **int** : **round** *a* ÷ **round** *b* }

7.0.2. Semantics

A mode-identifier (operator) which was caused to possess a value by the elaboration of a declaration containing the defining (operator-defining) occurrence

ADAPTED : ADJUSTED ; widened ; rowed ; hipped ; voided.
ADIC : PRIORITY ; monadic.
ADJUSTED : FITTED ; procedured ; united.
ALPHA : a ; b ; c ; d ; e ; f ; g ; h ; i ; j ; k ; l ; m ; n ; o ; p ; q ; r ; s ; t ; u ; v ; w ; x ; y ; z.
ANY : KIND ; suppressible KIND ; replicatable KIND ; replicatable suppressible KIND.
BITS : structured with row of boolean field LENGTHETY letter aleph.
BOX : LMOODSETY box.
BYTES : structured with row of character field LENGTHETY letter aleph.
CLAUSE : MOID clause.
CLOSED : closed ; collateral ; conditional.
COERCEND : MOID FORM.
COMPLEX : structured with real field letter r letter e and real field letter i letter m.
DIGIT : digit FIGURE.
EIGHT : SEVEN plus one.
EMPTY : .
FEAT : firm ; weak ; soft.
FIELD : MODE field TAG.
FIELDS : FIELD ; FIELDS and FIELD.
FIGURE : zero ; one ; two ; three ; four ; five ; six ; seven ; eight ; nine.
FITTED : dereferenced ; deprocedured.
FIVE : FOUR plus one.
FORESE : ADIC formula ; cohesion ; base.
FORM : confrontation ; FORESE.
FOUR : THREE plus one.
INTEGRAL : LONGSETY integral.
INTREAL : INTEGRAL ; REAL.
KIND : sign ; zero ; digit ; point ; exponent ; complex ; string ; character.
LENGTH : letter l letter o letter n letter g.
LENGTHETY : LENGTH LENGTHETY ; EMPTY.
LETTER : letter ALPHA ; letter aleph.
LFIELDSETY : FIELDS and ; EMPTY.
LIST : list ; sequence.
LMODE : MODE.
LMOOD : MOOD and.
LMOODS : LMOOD ; LMOODS LMOOD.
LMOODSETY : MOOD and LMOODSETY ; EMPTY.
LMOOT : MOOD and.
LONGSETY : long LONGSETY ; EMPTY.
LOSETY : LMOODSETY.
LOWPER : lower ; upper.
MABEL : MODE mode ; label.
MODE : MOOD ; UNITED.
MOID : MODE ; void.
MOOD : TYPE ; STOWED.
NINE : EIGHT plus one.
NONPROC : PLAIN ; format ; procedure with PARAMETERS MOID ; reference to NONPROC ; structured with FIELDS ; row of NONPROC ; UNITED.
NONROW : NONSTOWED ; structured with FIELDS.
NONSTOWED : TYPE ; UNITED.
NOTION : ALPHA ; NOTION ALPHA.
NUMBER : one ; TWO ; THREE ; FOUR ; FIVE ; SIX ; SEVEN ; EIGHT ; NINE.
PACK : pack ; package.
PARAMETER : MODE parameter.
PARAMETERS : PARAMETER ; PARAMETERS and PARAMETER.
PARAMETY : with PARAMETERS ; EMPTY.
PHRASE : declaration ; CLAUSE.
PLAIN : INTREAL ; boolean ; character.
PRAM : procedure with LMODE parameter and RMODE parameter MOID ; procedure with RMODE parameter MOID.
PRIMITIVE : integral ; real ; boolean ; character ; format.
PRIORITY : priority NUMBER.
PROCEDURE : procedure PARAMETY MOID.

REAL : LONGSETY real.
REFETY : reference to ; EMPTY.
RFIELDSETY : and FIELDS ; EMPTY.
RMODE : MODE.
RMOODSETY : RMOODSETY and MOOD ; EMPTY.
ROWS : row of ; ROWS row of.
ROWSETY : ROWS ; EMPTY.
ROWWSETY : ROWSETY.
SEPARATOR : LIST separator ; go on symbol ; completer ; sequencer.
SEVEN : SIX plus one.
SIX : FIVE plus one.
SOME : serial ; unitary ; CLOSED ; choice ; THELSE.
SORT : strong ; FEAT.
SORTETY : SORT ; EMPTY.
STIRM : strong ; firm.
STOWED : structured with FIELDS ; row of MODE.
STRING : row of character ; character.
STRONGETY : strong ; EMPTY.
TAG : LETTER ; TAG LETTER ; TAG DIGIT.
THELSE : then ; else.
THREE : TWO plus one.
TWO : one plus one.
TYPE : PLAIN ; format ; PROCEDURE ; reference to MODE.
UNITED : union of LMOODS MOOD mode.
VICTAL : VIRACT ; formal.
VIRACT : virtual ; actual.

of that mode-identifier (operator) is caused to possess an undefined value upon termination or completion of the elaboration of the smallest range {4.1.1.e} containing that declaration.

7.1. Declarers

{Declarers are built from the symbols **int**, **real**, **bool**, **char** and **format**, with the assistance of certain other symbols as, e.g., **long**, **ref**, [,], **struct**, **union** and **proc**. A declarer specifies a mode, e.g., **real** specifies the mode 'real'. A declarer is either a declarator or a mode-indication, e.g., **compl** is a mode-indication and not a declarator. Declarers are classified as actual, formal or virtual depending on the kind of lower- and upper-bounds which are permitted. Formal-declarers have the greatest freedom in this respect; e.g., [*1* : *n*] **real**, [*1* : *n* **flex**] **real**, [*1* : **flex**] **real**, [*1* : *n* **either**] **real**, [*1* : **either**] **real** and [ ] **real** may all be formal, but only the first two may be actual and only the last one may be virtual.}

7.1.1. Syntax

a)* declarer : VICTAL MODE declarer{b}.
b) VICTAL MODE declarer{h,k,l,m,n,o,p,x,y,aa,jj,54e,72a,834a,851b,c} :
  VICTAL MODE declarator{c,d,e,l,m,n,o,p,w,cc} ;
  MODE mode indication{42b}.
c) VICTAL PRIMITIVE declarator{b,d} : PRIMITIVE symbol{31d}.
d) VICTAL long INTREAL declarator{b,d} :
  long symbol{31d}, VICTAL INTREAL declarator{c,d}.

{Examples:
b) **real** ; **bits** ;
c) **int** ; **real** ; **bool** ; **char** ; **format** ;
d) **long int** ; **long long real** }

e) VICTAL structured with FIELDS declarator{b} :
  structure symbol{31d}, VICTAL FIELDS declarator{f,h,k} pack.
f) VICTAL FIELDS and FIELD declarator{e,f} :
  VICTAL FIELDS declarator{f,h,k}, comma symbol{31e},
   VICTAL FIELD declarator{h,k}.
g)* field declarator : VICTAL FIELD declarator{h,k}.
h) VICTAL STOWED field TAG declarator{e,f} :
  VICTAL STOWED declarer{b}, STOWED field TAG selector{j}.
i)* field selector : FIELD selector{j}.
j) MODE field TAG selector{h,k,852a} : TAG{302b,41c,d}.
k) VICTAL NONSTOWED field TAG declarator{e,f} :
  virtual NONSTOWED declarer{b}, NONSTOWED field TAG selector{j}.

{Examples:
e) **struct** *(***string** *title,* [*1* : *n*] **ref string** *pages,* **int** *price)* ;
f) **string** *title,* [*1* : *n*] **ref string** *pages,* **int** *price* ;
h) [*1* : *n*] **ref string** *pages* ;
j) *title* ;
k) **int** *price* }

{Rules h and k, together with 1.2.1.r,s,t,u,v and 4.1.1.c,d lead to an infinity of production rules of the strict language, thereby enabling the syntax to "transfer" the field-selectors (i) into the mode of structured values, and making it ungrammatical to use an "unknown" field-selector in a selection (8.5.2). Concerning the occurrence of a given field-selector more than once in a declarer, see 4.4.3, which implies that **struct** *(***real** *x,* **int** *x)* is not a (correct) declarer, whereas **struct** *(***real** *x,* **struct** *(***int** *x,* **bool** *p) p)* is. Notice, however, that the use of a given field-selector in two different declarers within a given reach does not cause ambiguity. Thus, **mode cell** = **struct** *(***string** *name,* **ref cell** *next)* and **mode link** = **struct** *(***ref link** *next,* **ref cell** *value)* may both occur in the same reach.}

l) VIRACT reference to MODE declarator{b} :
 reference to symbol{31d}, virtual MODE declarer{b}.
m) formal reference to STOWED declarator{b} :
 reference to symbol{31d}, formal STOWED declarer{b}.
n) formal reference to NONSTOWED declarator{b} :
 reference to symbol{31d}, virtual NONSTOWED declarer{b}.

{Examples:
l) **ref** [ ] **real** ;
m) **ref** [*1* :] **real** ; **ref** [*1* : **either**, *1* : **flex**] **real** ;
n) **ref ref** [ ] **real** }

{Rules l, m and n imply that, for instance, **ref** [*1* : **either**] **real** *x* may be a formal-parameter (5.4.1.e), whereas **ref ref** [*1* : **either**] **real** *x* may not.}

o) VICTAL ROWS structured with FIELDS declarator{b} :
 sub symbol{31e}, VICTAL ROWS rower{q,r}, bus symbol{31e},
  VICTAL structured with FIELDS declarer{b}.
p) VICTAL ROWS NONSTOWED declarator{b} :
 sub symbol{31e}, VICTAL ROWS rower{q,r}, bus symbol{31e},
  virtual NONSTOWED declarer{b}.
q) VICTAL row of ROWS rower{o,p,q} :
 VICTAL row of rower{r}, comma symbol{31e}, VICTAL ROWS rower{q,r}.
r) VICTAL row of rower{o,p,q} :
 VICTAL lower bound{s,t,v}, up to symbol{31e}, VICTAL upper bound{s,t,v}.
s) virtual LOWPER bound{r} : EMPTY.
t) actual LOWPER bound{r} :
 strict LOWPER bound{u}, flexible symbol{31d} option.
u) strict LOWPER bound{t,v,861f} : strong integral tertiary{81b}.
v) formal LOWPER bound{r} :
 strict LOWPER bound{u} option, flexible symbol{31d} option ;
 strict LOWPER bound{u} option, either symbol{31d}.

{Examples:
o) [*1* : *m*] **struct** *([1* : *n]* **real** *a,* **int** *b)* ;
p) [*1* : *m*, *1* : *n*] **ref** [ ] **real** ;
q) *1* : *m*, *1* : *n* ;
r) *1* : *m* ;

t) *m* ; *m* **flex** ;

u) *m* ;

v) *m* ; ; *m* **flex** ; **flex** ; *m* **either** ; **either** }

{The flexible-symbol, either-symbol, strict-lower-bound and strict-upper-bound contained in a formal-declarer serve to prescribe states and bounds of the multiple value possessed by the corresponding actual-parameter. The flexible-symbol in **ref** [*1* : **flex**] **char** *s* = *t* prescribes that a name referring to a multiple value with upper state *0* (i.e., the upper bound may vary) will be possessed by *s*; in **ref** [*1* : *n* **either**] **char** *s* = *t*, the either-symbol allows that upper state to be either *0* or *1* (i.e., the upper bound may be variable or fixed) and the absence of both flexible-symbol and either-symbol in **ref** [*1* : *n*] **char** *s* = *t* prescribes that that upper state is *1* (i.e., the upper bound must be fixed). Independently, the strict-upper-bound *n* in **ref** [*1* : *n* **either**] **char** *s* = *t* or in **ref** [*1* : *n*] **char** *s* = *t* prescribes that a name referring to a multiple value whose upper bound equals the value of *n* will be possessed by *s*; if, in the first example, the upper state is *0*, then that upper bound may well be changed later on by an assignment. The absence of a strict-upper-bound in **ref** [*1* : **flex**] **char** *s* = *t* does not restrict the upper bound in that way. Similar remarks apply to strict-lower-bounds. The flexible-symbol, strict-lower-bound and strict-upper-bound serve a similar role in generators (8.5).}

w) VICTAL PROCEDURE declarator{b} :
procedure symbol{31d}, virtual PROCEDURE plan{x,aa}.

x) virtual procedure with PARAMETERS MOID plan{w,75b} :
virtual PARAMETERS{y,54c} pack, virtual MOID declarer{b,z}.

y) virtual MODE parameter{x,54c} : virtual MODE declarer{b}.

z) virtual void declarer{x,aa,834a} : EMPTY.

aa) virtual procedure MOID plan{w} : virtual MOID declarer{b,z}.

bb)* parameters pack : VICTAL PARAMETERS{y,54c,e,74b} pack.

{Examples:

w) **proc** ; **proc** *(***real, int***)* ; **proc bool** ; **proc** *(***real***)* **bool** ;

x) *(***real, int***)* ; *(***real***)* **bool** ;

y) **real** ;

aa) **bool** }

cc) VICTAL union of LMOODS MOOD mode declarator{b} :
union of symbol{31d}, LMOODS MOOD and open box{dd} pack.

dd) LOSETY LMOOD open BOX{cc,ee} : LOSETY closed LMOOD end BOX{ee,ff}.

ee) LOSETY closed LMOODSETY LMOOD end BOX{dd,ee,ff} :
LOSETY closed LMOODSETY LMOOD LMOOD end BOX{ee,ff} ;
LOSETY open LMOODSETY LMOOD BOX{dd,gg}.

ff) LOSETY closed LMOODSETY LMOOD end LMOOT BOX{dd,ee,ff} :
LOSETY closed LMOODSETY LMOOT LMOOD end BOX{ee,ff}.

gg) open LMOODS LMOOD BOX{ee,gg,ii} : LMOODS LMOOD BOX{ii} ;
open LMOODS box{gg,hh}, comma symbol{31e}, LMOOD BOX{ii,jj}.

hh) open LMOOD box{gg} : LMOOD box{jj}.

ii) LMOODS MOOD and box{gg} : union of LMOODS MOODmode mode indication{42b} ;
union of symbol{31d}, open LMOODS MOOD and box{gg} pack.

jj) MOOD and box{gg,hh} : virtual MOOD declarer{b}.

{Examples:

cc) **union** *(***real**, **union** *(***int**, **bool***)*, **union** *(***real**, **int***))* ;
union *(***ri**, **union** *(***bool**, **real***))* (in the reach of **union ri** = *(***real**, **int***)*)}

{Let "*b*" stand for 'boolean', "*i*" for 'integral', "*r*" for 'real', "+" for 'and' and "*(bir)*" for any of the six protonotions '*b*+*i*+*r*','*b*+*r*+*i*','*i*+*b*+*r*', '*i*+*r*+*b*', '*r*+*b*+*i*' and '*r*+*i*+*b*'. Both examples are then examples of a virtual-, actual- or formal-union-of-*(bir)*-mode-declarator. The choice for *(bir)* is left undefined and is semantically irrelevant, but if one chooses some canonical ordering of all modes involved in a program, then the rules cc up to jj and 8.2.4.1.a,b,c,d do not cause any ambiguity (see 1.1.6.i, 4.4.3.d). The production mechanism of the rules cc up to jj is such that rule ee1 repeats, rule ff commutes and rule gg associates modes, whereas rule dd closes and rule ee2 opens the box. Let "#" stand for 'box', "(" for 'closed', ")" for 'end', "( )" for 'open' and "," for ', comma symbol,', then the production of the first example from 'actual union of integral and real and boolean mode declarator' is suggested by:

cc	*i*+*r*+*b*+( )#	ee1	*i*+(*r*+*r*+)*b*+#
dd	*i*+*r*+(*b*+)#	ff	*i*+(*r*+*b*+*r*+)#
ee2	*i*+*r*+( )*b*+#	ee2	*i*+( )*r*+*b*+*r*+#
dd	*i*+(*r*+)*b*+#	dd	(*i*+)*r*+*b*+*r*+#
ff	(*r*+*i*+)*b*+*r*+#	ee2	( )*r*+*i*+*b*+*r*+*i*+#
ee1	(*r*+*i*+*i*+)*b*+*r*+#	gg2	( )*r*+*i*+*b*+#,*r*+*i*+#
ff	(*r*+*i*+*b*+*i*+)*r*+#	gg2	( )*r*+#,*i*+*b*+#,*r*+*i*+#
ff	(*r*+*i*+*b*+*r*+*i*+)#	hh	*r*+#,*i*+*b*+#,*r*+*i*+#}

7.1.2. Semantics

a) A given declarer specifies the mode enveloped {1.1.6.j} by its original {1.1.6.c}.

b) A given declarer is protected by protecting all its constituent serial-clauses.

c) A given declarer is "developed" in the following steps:

Step 1: It is protected;

Step 2: If it is, or contains and does not shield, a mode-indication which is an actual-declarer or formal-declarer, then that indication is replaced by a copy of the protected actual-declarer of that mode-declaration {7.2} whose mode-indication is its indication-defining occurrence {4.2.2.a}, and Step 2 is taken; otherwise, the development of the declarer has been accomplished.

{A declarer is developed during the elaboration of an actual-declarer (d) or identity-declaration (7.4.2.Step 1).}

d) A {virtual- or actual-} declarer *D* specifying the mode *M* is elaborated in the following steps:

Step 1: If D is a virtual-declarer, then a new instance V of some value of some mode N {not beginning with 'union of' and} such that M is, or is united from, N is chosen and Step 8 is taken; otherwise, D is developed and Step 2 is taken;

Step 2: If D now begins with a structure-symbol, then Step 4 is taken; otherwise, if D now begins with a sub-symbol, then Step 5 is taken; otherwise, if D now begins with a union-of-symbol, then Step 3 is taken; otherwise, a new instance V of some value of the mode M is chosen and Step 8 is taken;

Step 3: Some mode N is chosen which does not begin with 'union of' and from which M is united {4.4.3.a}; a new instance V of some value of the mode N is chosen and Step 8 is taken;

Step 4: All constituent declarers of D are elaborated collaterally {6.2.2.a}; the values referred to by the values {names} of these declarers are made, in the given order, to be the fields of a new instance V of a structured value of the mode M, and Step 8 is taken;

Step 5: All constituent boundscripts {8.6.1.1.l} of D are elaborated collaterally;

Step 6: A descriptor {2.2.3.3} Q is established consisting of an offset 1 and as many quintuples, say n, as there are constituent actual-row-of-rowers in D; for $i=1, \ldots, n$, l_i (u_i) is set equal to the value of the constituent strict-lower-bound (strict-upper-bound) of the i-th of these actual-row-of-rowers; if the flexible-symbol-option of the actual-lower-bound (actual-upper-bound) of the i-th of these actual-row-of-rowers is a flexible-symbol, then s_i (t_i) is set to 0; otherwise, s_i (t_i) is set to 1; next, d_n is set to 1, and for $i=n, n-1, \ldots, 2$, the stride d_{i-1} is set to $(u_i-l_i+1)\times d_i$;

Step 7: Q is made to be the descriptor of a new instance V of a multiple value of the mode M, whose elements are obtained as follows: the last constituent declarer of D is elaborated collaterally a number of times and the elements are copies of values referred to by {some of the} resulting names;

Step 8: A name {2.2.3.5} different from all other names and whose mode is 'reference to' followed by M, is created and made to refer to V; this name is the value of D.

7.2. Mode Declarations

{Mode-declarations provide the indication-defining occurrences of mode-indications, which act as abbreviations for declarers built from some basic-tokens, e.g., **mode string** = [*1* : *0* **flex**] **char**, or from other declarers or even from themselves, e.g., **mode book** = **struct** *(***string** *title,* **ref book** *next).* In this last example, the mode-indication is not only a convenient abbreviation, but it is essential to the declaration.}

7.2.1. Syntax

a) mode declaration{70a} : mode symbol{31d}, MODE mode indication{42b}, equals symbol{31c}, actual MODE declarer{71b}.

{Examples:

a) **mode string** = [*1* : *0* **flex**] **char**; **struct compl** = *(***real** *re, im)*; **union primitive** = *(***int, real, bool, char, format***)* (see 9.2.b,c)}

7.2.2. Semantics

The elaboration of a mode-declaration involves no action.

{See 4.4.4.c concerning certain mode-declarations, e.g., **mode a** = **a**, which are not contained in proper programs.}

7.3. Priority Declarations

{Priority declarations provide the indication-defining occurrences of dyadic-indications, e.g., **o** in **priority o** = *6*, which may then be used in the declaration of dyadic operations. Priorities from *1* to *9* are available. Since monadic-operators have effectively only one priority level (8.4.1.g), which is higher than that of all dyadic-operators, monadic-indications do not occur in priority-declarations.}

7.3.1. Syntax

a) priority declaration{70a} :
   priority symbol{31d}, priority NUMBER indication{42e},
   equals symbol{31c}, NUMBER token{b,c,d,e,f,g,h,i,j}.

b) one token{a} : digit one symbol{31b}.

c) TWO token{a} : digit two symbol{31b}.

d) THREE token{a} : digit three symbol{31b}.

e) FOUR token{a} : digit four symbol{31b}.

f) FIVE token{a} : digit five symbol{31b}.

g) SIX token{a} : digit six symbol{31b}.

h) SEVEN token{a} : digit seven symbol{31b}.

i) EIGHT token{a} : digit eight symbol{31b}.

j) NINE token{a} : digit nine symbol{31b}.

{Example:

a) **priority** + = *6* }

7.3.2. Semantics

The elaboration of a priority-declaration involves no action.

{For a summary of the standard priority-declarations, see the remarks in 8.4.2.}

7.4. Identity Declarations

{Identity-declarations provide the defining occurrences of mode-identifiers, e.g., *x* in **real** *x* (which is an abbreviation of **ref real** *x* = **loc real** , see 9.2.a). Their elaboration causes mode-identifiers to possess values; here, *x* is made to possess a name which refers to some real number.}

7.4.1. Syntax

a) identity declaration{70a} : formal MODE parameter{54e},
   equals symbol{31c}, actual MODE parameter{b}.

b) actual MODE parameter{a,54c,75a,862a} :
   strong MODE unit{61e} ; MODE transformat{558a,—}.

{Examples:

a) **real** *e* = *2.7182818284590 45*; **int** *e* = **abs** *i*; **real** *d* = **re** (*z* × **conj** *z*);
ref [,] **real** *a1* = *a* [, : *k*]; **ref real** *x1k* = *x1* [*k*]; **compl** *unit* = *1*;
proc int *time* = *clock* ÷ *cycles*;
(The following declarations are given first without, and then with, the extensions of 9.2)
ref real *x* = **loc real**; **real** *x*;
ref int *sum* = **loc int** := *0*; **int** *sum* := *0*;
ref [**either** : **either**, **either** : **either**] **real** *a* = **loc** [*1* : *m*, *1* : *n*] **real** := *x2*;
[*1* : *m*, *1* : *n*] **real** *a* := *x2*;
proc (**real**) **real** *vers* = ((**real** *x*) **real** : *1* − *cos* (*x*));
proc *vers* = (**real** *x*) **real** : *1* − *cos* (*x*);
ref proc (**int**) **int** *q* = **loc proc** (**int**) **int** := ((**int** *i*) **int** : **abs** *i*);
proc *q* := (**int** *i*) **int** : **abs** *i*;

b) **abs** *i*; **loc real**; **loc int** := *0*; $+*d.11de*+*2d*$ }

7.4.2. Semantics

An identity-declaration is elaborated in the following steps:

Step 1: The formal-declarer *D* of its formal-parameter *F* is developed {7.1.2.c};

Step 2: Its actual-parameter *A* and all boundscripts contained in *D*, as possibly modified in Step 1, but not contained in any boundscript contained in *D*, are elaborated collaterally {6.2.2.a};

Step 3: If the value *V* of *A* is a name which refers to a component {2.2.2.k} of a multiple value having one or more states equal to *0*, then the further elaboration is undefined; otherwise, if *V* is a name other than *nil*, then the value to which *V* refers, or otherwise *V* itself, is termed *W*;

Step 4: If *W* is not a structured value or multiple value, then Step 7 is taken; otherwise, if *V* is not a name, then Step 6 is taken;

Step 5: For each flexible-symbol-option *S* contained in *D*, as possibly modified in Step 1, but not contained in any boundscript contained in *D*, {the corresponding state is checked, i.e.,} if *S* is a flexible-symbol (empty) and the corresponding state in *W* is *1* (*0*), then the further elaboration is undefined; otherwise, Step 6 is taken;

Step 6: For each boundscript contained in *D*, as possibly modified in Step 1, but not contained in any boundscript contained in *D* and not followed by a flexible-symbol, {the corresponding bound is checked, i.e.,} if its value is not the same as the corresponding bound, if any, in *W*, then the further elaboration is undefined; otherwise, Step 7 is taken;

Step 7: The identifier of *F* is made to possess *V*.

{According to Step 6, the elaboration of the declaration
[*1* : *2*] **real** *x1* = (*1.2*, *3.4*, *5.6*)
is undefined and according to Step 5 the elaboration of the declaration
ref [*1* : **flex**] **real** *x1* = [*1* : *2*] **real** := (*1.2*, *3.4*)
is undefined. The elaboration of the declaration
[*1* : **flex**] **real** *x1* = (*1.2*, *3.4*)

is well defined, but its effect is also obtained by the elaboration of the less confusing declaration

[ ] **real** *x1* = *(1.2, 3.4)*.}

7.5. Operation Declarations

{Operation-declarations provide the operator-defining occurrences of operators, e.g., **op** ∨ = *(***real** *a, b)* **real** : *(random* < *.5* | *a* | *b)*, which contains an operator-defining occurrence of ∨ as a dyadic-operator. Unlike identity-declarations of which no two for the same identifier may occur in a reach (4.4.2.b), more than one operation-declaration involving the same adic-indication may occur in the same reach, see 10.2.3.i, 10.2.4.i, etc.}

7.5.1. Syntax

a) operation declaration{70a} :
   PRAM caption{b}, equals symbol{31c}, actual PRAM parameter{74b}.

b) PRAM caption{a} : operation symbol{31d},
   virtual PRAM plan{71x}, PRAM ADIC operator{43b,c,—}.

{Examples:

a) **op** ∧ = *(***bool** *a, b)* **bool** : *(a* | *b* | **false***)* ;
   op abs = *(***real** *a)* **real** : *(a* < *0* | − *a* | *a)* (see 9.2.d,e) ;

b) **op** *(***bool, bool***)* **bool** ∧ ; **op** *(***real***)* **real abs** }

7.5.2. Semantics

An operation-declaration is elaborated in the following steps:

Step 1: Its actual-parameter is elaborated;

Step 2: The operator of its caption is made to possess the {routine which is the} value obtained in Step 1.

{The formula (8.4.1) *p* ∧ *q*, where ∧ identifies the operator-defining occurrence of ∧ in the operation-declaration

op ∧ = *(***bool** *john*, **proc bool** *mccarthy)* **bool** : *(john* | *mccarthy* | **false***)*,

possesses the same value as it would if ∧ identified the operator-defining occurrence of ∧ in the operation-declaration

op ∧ = *(***bool** *a, b)* **bool** : *(a* | *b* | **false***)*,

except, possibly, when the elaboration of *q* involves side effects on that of *p*.}

8. Unitary Clauses

{Unitary-clauses may occur as actual-parameters, e.g., *x* in *sin (x)*, as sources in assignations, e.g., *y* in *x* := *y*, in casts, especially in routine-denotations, e.g., *i* +:= *1* in *((***ref int** *i)* **int** : *i* +:= *1)*, or may be used to construct serial-clauses or collateral-clauses, e.g., *x* := *1* in *(x* := *1; y* := *2)* or in *(x* := *1, y* := *2)*. Unitary-clauses either are closed, collateral or conditional, or are coercends. There are four kinds of coercends: confrontations, e.g., *x* := *1*, formulas, e.g., *x* + *1*, cohesions, e.g., *next* **of** *cell*, and bases, e.g., *x*. These coercends and the closed-, collateral- and conditional-clauses are grouped into the following four classes, each class being a subclass of the next: primaries, which may be subscripted and parametrized, e.g., *x1* and *sin* in *x1* [*i*] and *sin(x)*; secondaries, from which fields may be selected, e.g., *z* in *re* **of** *z*; tertiaries, which may be operands, or may be destinations in assignations, or may occur in identity- or conformity-relations, e.g., *x* in *x* + *1* or in *x* := *1* or in *x* :=: *yy* or in *x* ::= *ir*, or may be boundscripts, e.g., *m*, *0* and *n* in *x2* [: *m* @ *0*, *n*], and, finally, unitary-clauses, which is the largest class. Thus, *r* **of** *s (i)* means that *s* is first called or subscripted, whereas *(r* **of** *s) (i)* means that the field is selected first. Also, *r* **of** *s* + *t* means that the field is selected from *s* before elaborating the routine possessed by +, while to force the elaboration of + first, one must write *r* **of** *(s* + *t)*.}

8.1.1. Syntax

a) SORTETY unitary MOID clause{61e} : SORTETY MOID tertiary{b} ;
SORTETY MOID confrontation{820d,e,f,g,830a,—}.

b) SORTETY MOID tertiary{a,71u,831b,832a,833a,861h,i} :
SORTETY MOID secondary{c} ;
SORTETY MOID ADIC formula{820d,e,f,g,84b,g}.

c) SORTETY MOID secondary{b,84f,852a} : SORTETY MOID primary{d} ;
SORTETY MOID cohesion{820d,e,f,g,850a}.

d) SORTETY MOID primary{c,861a,862a} :
SORTETY MOID base{820d,e,f,g,860a,b} ;
SORTETY CLOSED MOID clause{62b,c,d,f,63a,64a,—}.

{Examples:

a) *x* ; *x* := *1* ;

b) *x* ; *x* + *1* ;

c) *x* ; **real** ;

d) *x* ; *(x* + *1)* }

8.2. Coercends

{Coercends are of four kinds: bases, e.g., *x*, cohesions, e.g., *re* **of** *z*, formulas, e.g., *x* + *y* and confrontations, e.g., *x* := *1*. These are collectively considered as coercends because it is in their production rules that the basic coercions appear.

In current programming languages certain implicit changes of type are described, usually in the semantics. Thus *x* := *1* may mean that the integral value of *1* yields an equivalent real value which is then assigned to the name possessed by *x*. In ALGOL 68, such implicit changes of mode are known as coercions, and are reflected in the syntax. Certain coercions available in other languages, such as that in *i* := *x*, are not permitted. One must write *i* := **round** *x* or *i* := **entier** *x*, for in this situation it is felt advisable for the programmer to state the coercion explicitly. Apart from this, all the coercions which the programmer might reasonably expect are supplied.

There are eight basic coercions. They are: dereferencing, deproceduring, proceduring, uniting, widening, rowing, hipping and voiding. In *x* + *3.14*, the base *x*, whose a priori mode is 'reference to real', is dereferenced to 'real'; in *x* := *random*, the base *random*, whose a priori mode is 'procedure real', is deprocedured to 'real'; in **proc real** *p* = *x* + *3.14*, the formula *x* + *3.14*, whose a priori mode is 'real', is procedured to 'procedure real'; in **union** *(***int, real***)* *ir* := *1*, the base *1*, whose a priori mode is 'integral', is united to 'union of integral and real mode'; in *x* := *1*, the base *1*, whose a priori mode is 'integral', is widened to 'real'; in **string** *s* := *"a"*, the base *"a"*, whose a priori mode is 'character', is rowed to 'row of character'; in *x* := **skip**, the skip **skip**, which has no a priori mode, is hipped to 'real' and in *(x* := *1; y* := *2)*, the confrontation *x* := *1*, whose a priori mode is 'reference to real', is voided (, i.e., its value is ignored).

The kinds of coercion which are used depend upon three things: "syntactic position", a priori mode and a posteriori mode (, i.e., the modes before and after coercion). There are four sorts of syntactic position. They are: "strong" positions, i.e., actual-parameters, e.g., *x* in *sin (x)*, sources, e.g., *x* in *y* := *x*, conditions, e.g., *x*>*0* in *(x*>*0* | *x* | *0)*, subscripts, e.g., *i* in *x1* [*i*], etc.; "firm" positions, i.e., operands, e.g., *x* in *x* + *y*, transformats, e.g., *$5d$*, and certain primaries, e.g., *sin* in *sin (x)*; "weak" positions, i.e., certain primaries, e.g., *x1* in *x1* [*i*], and certain secondaries, e.g., *z* in *re* **of** *z*; and "soft" positions, i.e., destinations, e.g., *x* in *x* := *y*, and some other tertiaries, e.g., *xx* in *xx* :=: *x*.

Strong positions are so termed because the a posteriori mode is dictated entirely by the context. Such positions lead to the possibility of any of the eight basic coercions. Firm positions are, e.g., operands, in which widening, rowing, hipping and voiding must be excluded, since, otherwise, the identification of the operations involved in *i* + *j*, *x* + *y* (, supposing + to be declared also for 'row of real'), *i* + **skip** and *i* + *algol* could not be properly made. In the weak positions, only deproceduring and dereferencing are permitted, and special care must be taken that dereferencing removes a 'reference to' only if followed by 'reference to'. The *x1* in *x1* [*i*] := *1* demonstrates the necessity for this look-ahead. In the soft positions, the a posteriori mode is the a priori mode except for the removal of zero or more times 'procedure'. Thus in soft positions only deproceduring is performed.

In the productions of a notion, the sort (strong, firm, weak, soft) of position is passed on, or modified during balancing (to strong), and leads to basic coercions which appear in the production rules for coercends; moreover, the coercion must be completely expended in these rules. For example, *y* in *x* := *y* is a real-source and, therefore, a strong-real-unit (8.3.1.1.c); the sort 'strong' is passed through the productions of 'strong real unit' until a 'strong real base' is reached (8.1.1.d); this is then produced to 'strongly dereferenced to real base' (8.2.0.1.d), next to 'reference to real base' (8.2.1.1.a) and finally to 'reference to real mode identifier' (8.6.0.1.a).}

8.2.0.1. Syntax

a)* coercend : SORT COERCEND{d,e,f,g} ; SORTly ADAPTED to COERCEND {821a,b,822a,b,c,823a,824a,825a,b,c,d,826a,827a,828a,b,—}.

b)* SORT coercend : SORT COERCEND{d,e,f,g}.

c)* ADAPTED coercend : SORTly ADAPTED to COERCEND{821a,b,822a,b,c, 823a,824a,825a,b,c,d,826a,827a,828a,b,—}.

d) strong COERCEND{81a,b,c,d} : COERCEND{830a,84b,g,850a,860a,b,—} ;
strongly ADAPTED to COERCEND{821a,822a,823a,824a,825a,b,c,d,
826a,827a,828a,b,—}.

e) firm COERCEND{81a,b,c,d,84d,f} : COERCEND{830a,84b,g,850a,860a,b,—} ;
firmly ADJUSTED to COERCEND{821a,822a,823a,824a,—}.

f) weak COERCEND{81a,b,c,d} : COERCEND{830a,84b,g,850a,860a,b,—} ;
weakly FITTED to COERCEND{821b,822b,—}.

g) soft COERCEND{81a,b,c,d} : COERCEND{830a,84b,g,850a,860a,b,—} ;
softly deprocedured to COERCEND{822c}.

{Examples:

d) *3.14* (in *x* := *3.14*) ; *y* (in *x* := *y*) ;

e) *3.14* ; *x* (in *3.14* + *x*) ; *sin* (in *sin* (*x*)) ;

f) *x1* (in *x1* [*i*]) ; *zz* (in *re* **of** *zz* in the reach of **ref compl** *zz*) ;

g) *x* (in *x* := *1*) ; *x or y* (in *x or y* := *3.14*) }

8.2.1. Dereferenced Coercends

{Coercends are dereferenced when it is required that an initial 'reference to' should be removed from the a priori mode; e.g., in *x* := *y*, the a priori mode of *y* is 'reference to real' but the a posteriori mode required in this strong position is 'real'. Here, *y* possesses a name which refers to a real number and it is that real number which is assigned to (the name possessed by) *x*, not that name (possessed by *y*).}

8.2.1.1. Syntax

a) STIRMly dereferenced to MODE FORM{a,820d,e,822a,b,823a,824b,d,825a,b,c,
d,826a} : reference to MODE FORM{830a,84b,g,850a,860a} ;
STIRMly FITTED to reference to MODE FORM{a,822a}.

b) weakly dereferenced to reference to MODE FORM{b,820f} :
reference to reference to MODE FORM{830a,84b,g,850a,860a} ;
weakly FITTED to reference to reference to MODE FORM{b,822b}.

{Examples:

a) *y* (in *x* := *y* or in *x* + *y*) ; *yy* (in *x* := *yy* or in *x* + *yy*) ;

b) *rx1* (in *rx1* [*i*] in the reach of **ref** [ ] **real** *rx1*) }

8.2.1.2. Semantics

A dereferenced-coercend is elaborated in the following steps:

Step 1: It is preelaborated {1.1.6.i};

Step 2: If the value obtained in Step 1 is not *nil*, then the value of the dereferenced-coercend is a copy of the value referred to by the value {name} obtained in Step 1; otherwise, the further elaboration is undefined.

{Weak dereferencing must look ahead so that it does not remove a 'reference to' which precedes a mode which does not begin with 'reference to'. For example, in *x1* [*i*] := *y*, the primary *x1* should not be dereferenced, for *x1* [*i*] must possess a name. In *x1* [*i*] + *y*, the *x1* is not dereferenced but the base *x1* [*i*] is.}

8.2.2. Deprocedured Coercends

{Coercends are deprocedured when it is required that an initial 'procedure' should be removed from the a priori mode; e.g., in *x* := *random*, the a priori mode of *random* is 'procedure real' but the a posteriori mode required in this strong position is 'real'. Here, the routine possessed by *random* is elaborated and the real number yielded is assigned to (the name possessed by) *x*.}

8.2.2.1. Syntax

a) STIRMly deprocedured to MOID FORM{a,b,820d,e,821a,824b,d,825a,b,c,d, 826a,828b} : procedure MOID FORM{830a,84b,g,850a,860a} ; STIRMly FITTED to procedure MOID FORM{a,821a}.

b) weakly deprocedured to MODE FORM{820f,821b} : procedure MODE FORM{830a,84b,g,850a,860a} ; firmly FITTED to procedure MODE FORM{a,821a}.

c) softly deprocedured to MODE FORM{c,820g} : procedure MODE FORM{830a,84b,g,850a,860a} ; softly deprocedured to procedure MODE FORM{c}.

{Examples:

a) *random* (in *x* := *random* or in *x* + *random*) ;

b) *rz* (in *re* **of** *rz* in the reach of **proc compl** *rz* = **compl** : *(random, random)*) ;

c) *x or y* (in *x or y* := *1*) }

8.2.2.2. Semantics

A deprocedured-coercend is elaborated in the following steps:

Step 1: It is preelaborated {1.1.6.i};

Step 2: The deprocedured-coercend is replaced by a closed-clause which is a copy of {the routine which is} its prevalue obtained in Step 1, and the elaboration of that closed-clause is initiated; the value yielded, if any, is that of the deprocedured-coercend and if this elaboration is completed or terminated, then the closed-clause is replaced by the deprocedured-coercend before the elaboration of a successor is initiated.

{See also calls, 8.6.2.}

8.2.3. Procedured Coercends

{Coercends are procedured when it is required that an initial 'procedure' should be placed before the a priori mode (i.e., they should be turned into procedures without parameters), e.g., *x* := *1* in **proc real** *p* := *x* := *1*. Here, *1* is not assigned to *x*, but that routine which assigns *1* to *x* is assigned to *p*. Notice, that **proc** *p* := *x* := *1* is syntactically incorrect, since *x* := *1* must first be voided before it can be procedured to the mode 'procedure void'; the way to achieve this is by using a void-cast-pack: **proc** *p* := *(: x := 1)*. For the coercion in **proc** *stop* = *exit* see 8.2.7.}

8.2.3.1. Syntax

a) STIRMly procedured to procedure MOID FORM{a,820d,e,824b,826a} :
MOID FORM{830a,84b,g,850a,860a,b,—} ;
STIRMly dereferenced to MOID FORM{821a,—} ;
STIRMly procedured to MOID FORM{a,—} ;
STIRMly united to MOID FORM{824a,—} ;
STIRMly widened to MOID FORM{825a,b,c,d,—} ;
STIRMly rowed to MOID FORM{826a,—}.

{Examples:

a) *3.14* (in **proc real** *p* := *3.14*) ; *x* (in **proc real** *p* = *x*) ;
3.14 (in **proc proc real** *p* := *3.14*) ;
1 (in **proc union** *(***int, real***)* *p* := 1) ;
1 (in **proc real** *p* := *1*) ; *1* (in **proc** [ ] **int** *p* := *1*) }

8.2.3.2. Semantics

A procedured-coercend is elaborated in the following steps:

Step 1: A copy is made of it {itself, not its value}; if the mode enveloped by the original of the procedured-coercend is 'procedure' followed by a second mode {not by 'void'}, then the second mode is considered; otherwise, Step 3 is taken;

Step 2: A virtual-declarer, which, if it occurred in the smallest reach containing the procedured-coercend, would specify the considered mode, followed by a cast-of-symbol is placed before the copy;

Step 3: An open-symbol is placed before and a close-symbol is placed after the copy as possibly modified in Step 2; the thus modified copy is the {routine which is the} value of the procedured-coercend.

{The elaboration of the strong-procedure-real-base *(p | x1 | y1) [i]* yields the routine *(***real** : *(p | x1 | y1) [i])*, whereas that of the strong-conditional-procedure-real-clause *(p | x1[i] | y1[i])* yields either the routine *(***real** : *x1[i])* or the routine *(***real** : *y1[i])* depending on the value of *p*; similarly, the elaboration of the firm-procedure-real-confrontation *x* := *(a* := *a* + *1; y)* yields the routine *(***real** : *x* := *(a* := *a* + *1; y))*, whereas that of the firm-closed-procedure-real-clause *(a* := *a* + *1; x* := *y)* yields, apart from a change in the value of *a*, the routine *(***real** : *x* := *y)*; as last example, the elaboration of the strong-procedure-void-base *(: i* := *i* + *1)* yields the routine *((: i* := *i* + *1))*.}

8.2.4. United Coercends

{Coercends are united when it is required that the a priori mode should be changed to a mode united from (4.4.3.a) it, e.g., in **union** *(***int, real***)* *ir* := *2*, the base *2* is of the a priori mode 'integral', but the source of this assignation requires the mode 'union of integral and real mode'.}

8.2.4.1. Syntax

a) STIRMly united to union of LMOODS MOOD mode FORM{820d,e,823a,826a} :
one out of LMOODS MOOD mode FORM{b} ;
some of LMOODS MOOD and but not FORM{c}.

b) one out of LMOODSETY MOOD RMOODSETY mode FORM{a} :
MOOD FORM{830a,84b,g,850a,860a} ;
firmly FITTED to MOOD FORM{821a,822a} ;
firmly procedured to MOOD FORM{823a,—}.

c) some of LMOODSETY MOOD and RMOODSETY but not LOSETY FORM{a,c} :
some of LMOODSETY and MOOD RMOODSETY but not LOSETY FORM{c,d,—} ;
some of LMOODSETY RMOODSETY but not MOOD and LOSETY FORM{c,d,—}.

d) some of and LMOOD MOOD RMOODSETY but not LMOOT LOSETY FORM{c} :
union of LMOOD MOOD RMOODSETY mode FORM{830a,84b,g,850a, 860a} ;
firmly FITTED to union of LMOOD MOOD RMOODSETY mode FORM {821a,822a}.

{Examples:

a) *2* ; *ir* ;

b) *2* ; *i* ; **true** ;

d) *ri* ; *ir* (all in *(***union ir** = *(***int, real***);* **ir** *ir* := *2;* **ir** *ri* = *(p | i | x);*
union *(***ir, proc bool***) irb* := *ri; irb* := *ir; irb* := **true***)*)}

{In uniting, 'strong' leads to 'firm' in order that unions like that involved in **union** *(***int, real***) ir* := *1* should not cause ambiguities. In this example, if the base *1* is widened, then it cannot be united, i.e., in the order or productions in the syntax, uniting cannot be followed by widening.}

8.2.5. Widened Coercends

{Coercends are widened when it is required that the a priori mode should be changed from 'integral' to 'real' or from 'real' to 'COMPLEX', e.g., *1* in *z* := *1*, or from 'BITS' to 'row of boolean' or from 'BYTES' to 'row of character'.}

8.2.5.1. Syntax

a) strongly widened to LONGSETY real FORM{b,820d,823a,826a} :
LONGSETY integral FORM{830a,84b,g,850a,860a} ;
strongly FITTED to LONGSETY integral FORM{821a,822a}.

b) strongly widened to structured with REAL field letter r letter e and REAL field letter i letter m FORM{820d,823a,826a} :
REAL FORM{830a,84b,g,850a,860a} ;
strongly FITTED to REAL FORM{821a,822a} ;
strongly widened to REAL FORM{a}.

c) strongly widened to row of boolean FORM{820d,823a,826a} :
BITS FORM{830a,84b,g,850a,860a} ;
strongly FITTED to BITS FORM{821a,822a}.

d) strongly widened to row of character FORM{820d,823a,826a} :
BYTES FORM{830a,84b,g,850a,860a} ;
strongly FITTED to BYTES FORM{821a,822a}.

{Examples:
a) *1* (in *x* := *1*) ; *i* (in *x* := *i*) ;
b) *3.14* (in *z* := *3.14*) ; *x* (in *z* := *x*) ; *1* (in *z* := *1*) ;
c) **1 0 1** ; *t* (in [*1* : *3*] **bool** *b1* := (*p* | **1 0 1** | *t*)) ;
d) **ctb** *"abc"* ; *r* (in *s* := (*p* | **ctb** *"abc"* | *r*)) }

8.2.5.2. Semantics

A widened-coercend is elaborated in the following steps:

Step 1: It is preelaborated {1.1.6.i} and the value yielded is considered;

Step 2: If the considered value is an integer, then the real number equivalent to it {2.2.3.1.d} is considered instead; otherwise, if the considered value is a real number, then the structured {complex (10.2.7)} value composed of two fields, which are the considered value and the real number *0* of the same length number as that of the considered value and which are selected by letter-r-letter-e and letter-i-letter-m respectively is considered instead; otherwise, {the considered value is a structured value with one field and} the field of the considered value is considered instead;

Step 3: The value of the widened-coercend is a new instance of the considered value; its mode is that enveloped by the original of the widened-coercend.

{Widening may not be done in firm positions, for, otherwise, *x* := *i* + *1* might be ambiguous.}

8.2.6. Rowed Coercends

{Coercends are rowed when it is required that 'row of' should be placed either before the a priori mode or after an initial 'reference to' of the a priori mode; e.g., in [*1* : *1*] **real** *a1* := *3.14*, the a priori mode of the base *3.14* is 'real' but the a posteriori mode required in this strong position is 'row of real', whereas in **ref** [*1* :] **real** *a2* = *x*, the a priori mode of the base *x* is 'reference to real' but the a posteriori mode required is 'reference to row of real'. Here, the value to which *x* refers, is turned into a multiple value with a descriptor. Note that the value of *a2* [*1*] :=: *x* is *true*.}

8.2.6.1. Syntax

a) strongly rowed to REFETY row of MODE FORM{a,820d,823a} :
   REFETY MODE FORM{830a,84b,g,850a,860a} ;
   strongly ADJUSTED to REFETY MODE FORM{821a,822a,823a,824a,—} ;
   strongly widened to REFETY MODE FORM{825a,b,c,d,—} ;
   strongly rowed to REFETY MODE FORM{a,—} ;
   REFETY row of MODE FORM vacuum{b,—}.

b) row of NONROW base vacuum{a} : EMPTY.

{Examples:
a) *3.14* (in [*1* : *1*] **real** *x1* := *3.14*) ; *y* (in **ref** [*1* : *1*] **real** *x1* = *y*) ;
   3.14 (in [*1* : *1*] **proc real** *p* := *3.14*) ;
   3.14 (in [*1* : *1*] **compl** *z1* := *3.14*) ;
   3.14 (in [*1* : *1*, *1* : *1*] **real** *x2* := *3.14*) ;
      y (in **ref** [*1* : *1*, *1* : *1*] **real** *x2* = *y*) ;
      (the EMPTY following := in [*1* : *0*] **real** :=) }

8.2.6.2. Semantics

A rowed-coercend is elaborated in the following steps:

Step 1: The mode enveloped by the original of the rowed-coercend is considered; if that mode begins with 'row of row of' or if the rowed-coercend is not empty, then it is preelaborated {1.1.6.i}, the value obtained and its scope are considered and Step 3 is taken;

Step 2: A new instance of a multiple value {2.2.3.3} composed of zero elements and a descriptor consisting of an offset *1* and one quintuple (*1*, *0*, *1*, *1*, *1*) is considered and Step 7 is taken;

Step 3: If the considered mode does not begin with 'reference to', then Step 5 is taken; otherwise, if the considered value is not *nil*, then Step 4 is taken; otherwise, the elaboration of the rowed-coercend is complete, its value is a new instance of *nil* whose mode is the considered mode;

Step 4: That instance of the value to which the {name which is the} considered value refers is considered instead; if the considered value is a multiple value having one or more states equal to *0*, or if it is a component {2.2.2.k} of such a multiple value, then the further elaboration is undefined; otherwise, Step 5 is taken;

Step 5: If the considered value is a multiple value, then Step 6 is taken; otherwise, the instance of a multiple value composed of the considered value as only element and of a descriptor consisting of an offset *1* and one quintuple (*1*, *1*, *1*, *1*, *1*) is considered instead, and Step 7 is taken;

Step 6: Let *d* stand for $(u_1 - l_1 + 1) \times d_1$; the instance of a new multiple value, composed of the elements of the considered value and of a descriptor which is a copy of the descriptor of the considered value into which the additional quintuple (*1*, *1*, *d*, *1*, *1*) is inserted before the first quintuple, and in which all states have been set to *1*, is considered instead;

Step 7: If the considered mode does not begin with 'reference to', then the value of the rowed-coercend is the considered value; otherwise, a name *N* is made to refer to the considered value {whose scope is the prescope obtained in Step 1, and} whose mode is the considered mode; this name *N* is the value of the rowed-coercend.

8.2.7. Hipped Coercends

{Coercends are hipped when they are skips, jumps or nihils. Though there is no a priori mode, whatever mode is required by the context, is adopted; e.g., in **real** $x =$ **skip**, the base, **skip**, which has no a priori mode, is hipped to 'real'. Since hipped-coercends are so very accommodating, no other coercions may follow them (in the elaboration order); otherwise, ambiguities might appear. Consider, for example, the several meanings of the assignation **union** *(***int**, **real**, **bool**, **char***)* $u :=$ **skip**, supposing uniting could follow hipping.}

8.2.7.1. Syntax

a) strongly hipped to MOID base{820d} :
  MOID skip{b} ; MOID jump{c} ; MOID nihil{d,–}.

b) MOID skip{a} : skip symbol{31g}.
c) MOID jump{a} : go to symbol{31f} option, label identifier{41b}.
d) reference to MODE nihil{a} : nil symbol{31g}.

{Examples:
a) **skip** ; **go to** *grenoble* ; **nil** ;
b) **skip** ;
c) **go to** *grenoble* ; *st pierre de chartreuse* ;
d) **nil** }

8.2.7.2. Semantics

a) A skip is elaborated in the following steps:

Step 1: If the terminal production of the metanotion 'MOID' enveloped {1.1.6.j} by the original {1.1.6.c} of the skip is a mode, then this mode is considered and Step 2 is taken; otherwise, {it is 'void' and} the elaboration of the skip is complete;

Step 2: If the considered mode begins with 'union of', then some mode from which it is united {4.4.3.a} is considered instead;

Step 3: The value of the skip is a new instance of some value of the considered mode and whose scope is the program.

b) A jump is elaborated in the following steps:

Step 1: If the original of the jump envelops a mode which is 'procedure MOID' where "MOID" stands for any terminal production of the metanotion 'MOID', then this mode is considered and Step 2 is taken; otherwise, the elaboration of the jump is terminated and it appoints as its successor the unitary-clause following the label-sequence or the completer containing the defining occurrence {in a label (4.1.2)} identified by the label-identifier of the jump;

Step 2: A copy is made of the jump and an open-symbol followed by a cast-of-symbol is placed before and a close-symbol is placed after the copy; if the considered mode is not 'procedure void', then the initial 'procedure' is deleted from it and a virtual-declarer, which, if it occurred in the smallest reach containing the jump, would specify the mode so obtained, is inserted between the open-symbol and the cast-of-symbol in the copy; otherwise, an open-symbol is placed before and a close-symbol is placed after the thus modified copy;

Step 3: The value of the jump is the routine consisting of the same sequence of symbols as the copy as modified in Step 2 and whose mode is that enveloped by the original of the jump.

c) The elaboration of a nihil involves no action; its value is a new instance of *nil* {2.2.3.5.a} whose mode is that enveloped by the original of the nihil.

{Skips play a role in the semantics of routine-denotations (5.4.2.Step 2) and calls (8.6.2.2.Step 4). Moreover, they are useful in a number of programming situations, like e.g.,

supplying an actual-parameter (7.4.1.b) whose value is irrelevant or is to be calculated later; e.g., *f (3, ~)* where *f* does not use its second actual-parameter if the value of the first actual-parameter is positive; see also 11.11.ar;

supplying a constituent unit of a collateral-clause (6.2.1.b,c,d,f), e.g.,
[1 : 4] **real** *x1 := (3.14,* **skip**, *1.68,* **skip***)* ;

as a dummy statement (6.0.1.c) in those rare situations where the use of a completer is inappropriate, e.g., *l:* **skip***)* in 10.4.a. See also 9.4.a.

A jump is useful as a clause to terminate the elaboration of another clause when certain requirements are not met, e.g., **go to** *exit* in *y :=* **if** *x≥0* **then** *sqrt (x)* **else go to** *exit* **fi**.

If *e1*, *e2* and *e3* are label-identifiers, then the reader might recognize the effect of the declaration [ ] **proc** *switch = (e1, e2, e3)* and the statement *switch [i]*; however, the declaration *[1 : 3* **flex**] **proc** *switch := (e1, e2, e3)* is perhaps more powerful, since assignations like *switch [2] := e1* and *switch := (e1, e2, e3, e4)* are possible.

A nihil is particularly useful where structured values are connected to one another in that a field of each structured value refers to another one except for one or more structured values where the field does not refer to anything at all; such a field must then be *nil*.}

8.2.8. Voided Coercends

{Coercends are voided when it is required that their values (and therefore modes) should be ignored, e.g., in *(x := 1; y := 2)*, the confrontation *x := 1*, whose a priori mode is 'reference to real', is voided (see 6.1.1.i). Confrontations must be treated differently from the other coercends in order that, e.g., in *(***proc** *p; p := stop; p)*, the confrontation *p := stop* does not involve the elaboration of the routine possessed by *stop*, but in the last occurrence of *p*, that routine is elaborated.}

8.2.8.1. Syntax

a) strongly voided to void confrontation{820d} : MODE confrontation{830a}.

b) strongly voided to void FORESE{820d} :
NONPROC FORESE{84b,g,850a,860a} ;
strongly deprocedured to NONPROC FORESE{822a}.

{Examples:

a) *x := 1* (in *(x := 1; y := 2)*) ;

b) *x* ; *random* (in *(x; random;* **skip***)*) }

{The value obtained by elaborating (i.e., preelaborating 1.1.6.i) a voided-coercend is discarded.}

{In the reach of the declaration [ ] **proc** *switch = (e1, e2, e3)* and the clause-train *e1: e2: e3: stop*, the construction *switch; stop* is not a serial-clause because *switch* is not a strong-void-unit. In fact, *switch* can not be deprocedured, because its mode begins with 'row of' and no coercion will remove the 'row of' and it cannot be voided because 'row of procedure void' is not a terminal production of 'NONPROC'. However, the elaboration of *switch [2];* **skip** will involve a jump to the label *e2:*.}

8.3. Confrontations

8.3.0.1. Syntax

a) MODE confrontation{81a,820d,e,f,g,821a,b,822a,b,c,823a,824b,d,825a,b,c,d, 826a,828a} : MODE assignation{831a,—} ; MODE conformity relation {832a,—} ; MODE identity relation{833a,—} ; MODE cast{834a}.

{Examples:
a) *x* := *3.14* ; *ec* ::= *a* (see 11.11.i) ; *xx* :=: *x or y* ; [ ] **real** : *1* }

8.3.1. Assignations

{In assignations, e.g., *x* := *3.14*, (an instance of) a value is assigned to a name. In *x* := *3.14*, the value possessed by the source *3.14* is assigned to the (name which is the) value possessed by *x*.}

8.3.1.1. Syntax

a) reference to MODE assignation{830a} : reference to MODE destination{b}, becomes symbol{31c}, MODE source{c}.
b) reference to MODE destination{a} : soft reference to MODE tertiary{81b}.
c) MODE source{a} : strong MODE unit{61e}.

{Examples:
a) *x* := 1 ; **loc real** := *3.14* ;
b) *x* ; **loc real** ;
c) *1* ; *3.14* }

8.3.1.2. Semantics

a) When a given instance of a value is "superseded" by another instance of a value, then the name which refers to the given instance is caused to refer to that other instance, and, moreover, each name which refers to an instance of a structured or multiple value of which the given instance is a component {2.2.2.k} is caused to refer to the instance of the structured or multiple value which is established by replacing that component by that other instance.

b) When a field (an element) of a given structured (multiple) value is superseded by another instance of a value, then the mode of the thereby established structured (multiple) value is that of the given value.

c) An instance of a value is assigned to a name in the following steps:

Step 1: If the given value does not refer to a component of a multiple value having one or more states equal to *0* {2.2.3.3.b}, if the scope of the given name is not larger than the scope of the given value {2.2.4.2} and if the given name is not *nil*, then Step 2 is taken; otherwise, the further elaboration is undefined;

Step 2: The instance of the value referred to by the given name is considered; if the mode of the given name begins with 'reference to structured with' or with 'reference to row of', then Step 3 is taken; otherwise, the considered instance is superseded {a} by a copy of the given instance and the assignment has been accomplished;

Step 3: If the considered value is a structured value, then Step 5 is taken; otherwise, applying the notation of 2.2.3.3.b to its descriptor, if for some i, $i=1, \ldots, n$, $s_i=1$ ($t_i=1$) and l_i (u_i) is not equal to the corresponding bound in the descriptor of the given value, then the further elaboration is undefined;

Step 4: If some $s_i=0$ or $t_i=0$, then, first, a new instance of a multiple value M is created whose descriptor is a copy of the descriptor of the given value modified by setting its states to the corresponding states in the descriptor of the considered value, and whose elements are copies of elements, if any, of the considered value, and, otherwise, are new instances of values whose mode is, or is a mode from which is united, the mode obtained by deleting all initial 'row of's from the mode of the considered value; next, M is made to be referred to by the given name and is considered instead;

Step 5: Each field (element, if any,) of the given value is assigned {in an order which is left undefined} to the name referring to the corresponding field (element, if any,) of the considered value and the assignment has been accomplished.

d) An assignation is elaborated in the following steps:

Step 1: Its destination and source are elaborated collaterally {6.2.2.a};

Step 2: The value of its source is assigned to the {name which is the} value of its destination;

Step 3: The value of the assignation is the value of its destination.

{Observe that *(x, y) := (1.2, 3.4)* is not an assignation, since *(x, y)* is not a destination; the mode of the value of a collateral-clause (6.2.1.c,d,f) does not begin with 'reference to' but with 'row of' or 'structured with'.}

8.3.2. Conformity Relations

{The purpose of conformity-relations is to enable the programmer to find out the current mode of an instance of a value if the context permits this mode to be one of a number of given modes. See, for example, 11.11.i,q,y,ah. Conformity-relations are thus used in conjunction with unions.}

{*I would to God they would either conform, or be more wise, and not be catched!*
Diary, 7 Aug. 1664, *Samuel Pepys.*}

8.3.2.1. Syntax

a) boolean conformity relation{830a} :
soft reference to LMODE tertiary{81b}, conformity relator{b}, RMODE tertiary{81b}.

b) conformity relator{a} :
conforms to symbol{31c} ; conforms to and becomes symbol{31c}.

{Examples:

a) **int** :: *irb* ; *ec* ::= *a* (see 11.11.i) ;

b) :: ; ::= }

8.3.2.2. Semantics

A conformity-relation is elaborated in the following steps:

Step 1: Its textually last tertiary is elaborated and the value yielded is considered;

Step 2: If the mode enveloped by the original of its textually first tertiary is 'reference to' followed by a mode which is, or is united from {4.4.3.a}, the mode of the considered value, then the value of the conformity-relation is *true* and Step 4 is taken; otherwise, Step 3 is taken;

Step 3: If the considered value refers to another value, then this other value is considered instead and Step 2 is taken; otherwise, the elaboration is complete and the value of the conformity-relation is *false*;

Step 4: If its conformity-relator is a conforms-to-and-becomes-symbol, then its textually first tertiary is elaborated and the considered value is assigned {8.3.1.2.c} to the value of that tertiary.

{Although not suggested by the wording of Step 2, the, possibly, most obvious applications of conformity-relations are those in which 'RMODE' in 8.3.2.1.a begins with 'union of' whereas 'LMODE' does not. Then, the mode of the considered value (Step 1) is not 'RMODE' (which is united from it) and the conformity-relation serves to ask whether this mode is 'LMODE' and, if so and if the conformity-relator is a conforms-to-and-becomes-symbol, to assign this value to a name whose mode does not begin with 'reference to union of' and, thereby, make this value easily available elsewhere. Several applications, partly disguised by the application of the extensions 9.4 are given in 11.11.

Observe that if the considered value is an integer and the mode of its textually first tertiary is 'reference to' followed by a mode which is, or is united from, the mode 'real' but not from 'integral', then the value of the conformity-relation is *false*. Thus, no automatic widening from 'integral' to 'real' takes place. For example, in **union** (**real**, **bool**) *rb; rb* ::= *1*, no value is assigned to *rb*, but in *rb* ::= *1.0*, the assignment takes place. Rule 8.3.2.1.a is the only rule in the syntax where a notion other than a coercend produces uncoerced clauses, i.e., those produced from 'RMODE tertiary'.}

8.3.3. Identity Relations

{Identity-relations may be used to ask whether two names of the same mode are the same; e.g., in the reach of the declarations **struct cons** = (**ref cong** *car, cdr*); **union cong** = (**cons**, **string**); **cons** *cell* := (**cong** := *"abc"*, **nil**), the identity-relation *cdr* **of** *cell* :=: **nil** possesses the value *false* because the value of *cdr* **of** *cell* is the name referring to the second field of the structured value referred to by the value of *cell* and, hence, is not *nil*, but the value of (**ref cong** : *cdr* **of** *cell*) :=: **nil** is *true*.}

8.3.3.1. Syntax

a) boolean identity relation{830a} :
   soft reference to MODE tertiary{81b}, identity relator{b},
      strong reference to MODE tertiary{81b} ;
   strong reference to MODE tertiary{81b}, identity relator{b},
      soft reference to MODE tertiary{81b}.

b) identity relator{a} : is symbol{31c} ; is not symbol{31c}.

{Examples:

a) *x or y* :=: *x* ; *xx* :=: *x* ;

b) :=: ; :≠: }

8.3.3.2. Semantics

An identity-relation is elaborated in the following Steps:

Step 1: Its tertiaries are elaborated collaterally {6.2.2.a};

Step 2: If its identity-relator is an is-symbol (is-not-symbol), then the value of the identity-relation is ***true*** (***false***) if the {names which are the} values obtained in Step 1 are the same and ***false*** (***true***) otherwise.

{Assuming the assignation *xx* := *yy* := *x* to have been elaborated, the value of the identity-relation *xx* :=: *yy* is ***false*** because *xx* and *yy*, though of the same mode, do not possess the same name (7.1.2.Step 8), but the name which each possesses refers to the same name and so *(***ref real*** : xx)* :=: *(***ref real*** : yy)* possesses the value ***true***. The value of the identity-relation *xx* :=: *x or y* has a probability ½ of being ***true*** because the value possessed by *xx* (effectively that of **ref real** : *xx* here, because of coercion) is the name possessed by *x*, and the routine possessed by *x or y* (see 1.3), when elaborated, yields either the name possessed by *x* or, with equal probability, the name possessed by *y*.

In the identity-relation, the programmer is usually asking a specific question concerning names and thus the level of reference is of crucial importance. Thus at least one of the tertiaries of an identity-relation must be soft, i.e., must involve only deproceduring and certainly no dereferencing. The construction **case** *i* **in** *x, xx, x or y,* **nil esac** :=: **case** *j* **in** *y,* **skip**, *x or y, re* **of** *z* **out** *yy* **esac** is an example of a delicately balanced identity-relation in which the mode is 'reference to real'.

Observe that the value of the formula *1* = *2* is ***false***, whereas *1* :=: *2* is not an identity-relation, since the values of its tertiaries are not names. Also *$2d3d$* :=: *$5d$* is not an identity-relation, whereas *$2d3d$* = *$5d$* is a formula, but involves an operation which is not included in the standard-prelude.}

8.3.4. Casts

{Casts may be used to provide a strong position for a unitary-clause in a position which is not strong, e.g., **ref real** : *xx* in *(***ref real*** : xx)* := *1*. They play a role in routine-denotations (5.4.1.a), e.g., **real** : *a* + *1* in *((***int*** a)* **real** : *a* + *1)* and procedured-coercends (8.2.3.1.a), e.g., : *(l: l)* in **proc** *busy* = *(: (l: l))*. A void-cast is not a clause but is a constituent of a void-cast-pack and of some routine-denotations and thus of bases.}

8.3.4.1. Syntax

a) MOID cast{54b,830a,860b} : virtual MOID declarer{71b,z}, cast of symbol{31b}, strong MOID unit{61e}.

{Examples:

a) [ ] **real** : *1* ; : *x* := *3.14* }

8.3.4.2. Semantics

The elaboration (value, if any,) of a cast is that of its unit.

8.4. Formulas

{Formulas are either dyadic-formulas, e.g., $x+i$, or monadic-formulas, e.g., **abs** x. A formula contains at least one operand and at least one operator. The order of elaboration of a formula is determined by the priority of its operators; monadic-formulas are elaborated first and then the dyadic-formulas from the highest to the lowest priority.}

8.4.1. Syntax

a)⋆ SORTETY formula : SORTETY MOID ADIC formula{b,g,820d,e,f,g}.

b) MOID PRIORITY formula{81b,820d,e,f,g,821a,b,822a,b,c,823a,824b,d,825a,b, c,d,826a,828b} : LMODE PRIORITY operand{d}, procedure with LMODE parameter and RMODE parameter MOID PRIORITY operator{43b}, RMODE PRIORITY plus one operand{d,e}.

c)⋆ operand : MODE ADIC operand{d,f}.

d) MODE PRIORITY operand{b,d} : firm MODE PRIORITY formula{820e} ; MODE PRIORITY plus one operand{d,e}.

e) MODE priority NINE plus one operand{b,d} : MODE monadic operand{f}.

f) MODE monadic operand{e,g} : firm MODE monadic formula{820e} ; firm MODE secondary{81c}.

g) MOID monadic formula{81b,820d,e,f,g,821a,b,822a,b,c,823a,824b,d,825a,b, c,d,826a,828b} : procedure with RMODE parameter MOID monadic operator{43c}, RMODE monadic operand{f}.

h)⋆ dyadic formula : MOID PRIORITY formula{b}.

{Examples:

b) $x+x\times y$;

d) $x\times y$; x;

e) x;

f) **abs** x; *age* **of** *algol*;

g) − **abs re** z }

8.4.2. Semantics

A formula is elaborated in the following steps:

Step 1: The formula is replaced by a closed-clause which is a copy of the routine possessed by the operator-defining occurrence identified by its operator {7.5.2, 4.3.2.b};

Step 2: The constituent serial-clause of the closed-clause is protected {6.0.2.d};

Step 3: The skip-symbol {5.4.2.Step 2} following the equals-symbol following its textually first copied formal-parameter is replaced by a copy of the textually first operand of the formula, and if the formula is a dyadic-formula, then the skip-symbol following the equals-symbol following its textually second copied formal-parameter is replaced by a copy of the textually second operand of the formula;

Step 4: The closed-clause as modified in Steps 2 and 3 is replaced by a closed-clause consisting of the same sequence of symbols; the elaboration of this closed-clause is initiated; its value, if any, is then that of the formula and if this elaboration is completed or terminated, then this closed-clause is replaced by the formula before the elaboration of a successor is initiated.

{The following table summarises the priorities of the operators declared in the standard priorities (10.2.0).

dyadic									monadic
1	*2*	*3*	*4*	*5*	*6*	*7*	*8*	*9*	*(10)*
−:=	∨	∧	=	<	−	×	↑	⊥	¬ − + / ↓ ↑
+:=			≠	≤	+	÷	⌊		**abs** **bin** **repr**
×:=				≥		÷:	⌈		⌊ ⌈ ⎩ ⎧
/:=				>		/	⎩		**leng** **short**
÷:=						▯	⎧		**odd** **sign** **round**
÷::=									**re** **im** **conj**
+=:									**btb** **ctb**

Observe that $a \uparrow b$ is not precisely the same as a^b in usual notation; indeed, the value of $(-1 \uparrow 2+4=5)$ and that of $(4-1 \uparrow 2=3)$ both are *true*, since the first minus-symbol is a monadic-operator, whereas the second is a dyadic-operator. Although the syntax determines the order in which formulas are elaborated, parentheses may well be used to improve readability; e.g., $(a \wedge b) \vee (\neg a \wedge \neg b)$ instead of $a \wedge b \vee \neg a \wedge \neg b$.

In the formula $x+y\times 2$, both y and 2 are primaries, which allows y to be a firm-priority-SEVEN-operand and 2 to be a firm-priority-EIGHT-operand. The formula $y\times 2$ is then of priority **7**. Since x is also a primary, and therefore a firm-priority-SIX-operand, $x+y\times 2$ is a priority-SIX-formula. The effect of $x+y\times 2$ is thus the same as that of $x+(y\times 2)$.}

8.5. Cohesions

{Cohesions are of two kinds: generators, e.g., **string**, or selections, e.g., *re* **of** *z*. Cohesions are distinct from bases in order that constructions like *a* **of** *b* [*i*] may be parsed without knowing the mode of *a* and *b*. Cohesions may not be subscripted or parametrized, but they may be selected from, e.g., *father* **of** *algol* in *father* **of** *father* **of** *algol*.}

8.5.0.1. Syntax

a) MODE cohesion{81c,820d,e,f,g,821a,b,822a,b,c,823a,824b,d,825a,b,c,d, 826a,828b} : MODE generator{851a} ; MODE selection{852a}.

{Examples:

a) **real** (in *xx* := **real** := *3.14*) ; *re* **of** *z* }

8.5.1. Generators

{And as imagination bodies forth
The forms of things unknown, the poet's pen
Turns them to shapes, and gives to airy nothing
A local habitation and a name.
A Midsummer-night's Dream,
William Shakespeare.}

{The elaboration of a generator, e.g., **real** in *xx* := **real** := *3.14* or **loc real** in **ref real** *x* = **loc real** (usually written **real** *x* by extension 9.2.a), involves the creation of a name, i.e., the reservation of storage. The use of a local-generator implies (with most implementations) the reservation of storage in a run-time stack, whereas global-generators imply the reservation of storage in another region, termed the "heap", in which garbage-collection techniques may be used for storage retrieval. Since this is usually less efficient, global-generators should be avoided where possible. The temptation to use global-generators unnecessarily, is reduced by the extensions 9.2.a, which allow the greatest shortening of the text when local-generators are used.}

8.5.1.1. Syntax

a) MODE generator{850a} :
   MODE local generator{b,—} ; MODE global generator{c,—}.

b) reference to MODE local generator{a} :
   local symbol{31d}, actual MODE declarer{71b}.

c) reference to MODE global generator{a} :
   heap symbol{31d} option, actual MODE declarer{71b}.

{Examples:

a) **loc real** ; **heap real** ;

b) **loc real** ;

c) **heap real** ; **real** }

8.5.1.2. Semantics

a) A generator is elaborated in the following steps:

Step 1: Its actual-declarer is elaborated {7.1.2.d};

Step 2: The value of the generator is the {name which is the} value obtained in Step 1.

b) The scope {2.2.4.2} of the value of a local-generator is the smallest range containing that generator; that of a global-generator is the program.

{The closed-clause

*(***ref real** *xx; xx* := *(***heap real** *x* := *pi; x); xx* = *pi)* (see also 9.2.a)

possesses the value *true*, but the closed-clause

*(***ref real** *xx; xx* := *(***real** *x* := *pi; x); xx* = *pi)*

possesses an undefined value since the name to be assigned to the name possessed by *xx* becomes undefined upon the completion of the elaboration of the inner range, which is the scope of the name possessed by *x* (6.1.2.e, 7.0.2). The closed-clause

*((***ref real** *xx;* **real** *x := pi; xx := x) = pi)*

however, possesses the value *true*.}

8.5.2. Selections

{A selection selects a field from a structured value; e.g., *re* **of** *z* selects the first real field (usually termed the real part) of the value possessed by *z*. If *z* possesses a name, then *re* **of** *z* possesses also a name, but if *w* possesses a complex value, then *re* **of** *w* possesses a real value, not the name referring to a real value.}

8.5.2.1. Syntax

a) REFETY MODE selection{850a} : MODE field TAG selector{71j}, of symbol{31e}, weak REFETY structured with LFIELDSETY MODE field TAG RFIELDSETY secondary{81c}.

{Examples, assumed in the reach of the declarations:

struct language = *(***int** *age,* **ref language** *father);*

language *algol* := *(10,* **language** := *(14, ∾));* **language** *pl1* = *(4, algol);* :

a) *age* **of** *pl1* ; *father* **of** *algol* }

{Rule a ensures that the value of the secondary has a field selected by the field-selector in the selection (see 7.1.1.e,f,h,k and the remarks below 7.1.1 and 8.5.2.2). An identifier which is the same sequence of symbols as a field-selector in one same reach creates no ambiguity. Thus, *age* **of** *algol* := *age* is a (possibly confusing to the human) assignation if the second occurrence of *age* is an integral-mode-identifier.}

8.5.2.2. Semantics

A selection is elaborated in the following steps:

Step 1: Its secondary is elaborated; if its value is *nil*, then the further elaboration is undefined; otherwise, the structured value which is, or is referred to by, that value is considered;

Step 2: If the value of the secondary is a name, then the value of the selection is a new instance of the name which refers to that field of the considered structured value selected by its field-selector; otherwise, it is a new instance of {the value which is} that field itself.

{In the examples of 8.5.2.1, *age* **of** *algol* is a reference-to-integral-selection, and, by 8.5.0.1.a, a reference-to-integral-cohesion, but *age* **of** *pl1* is an integral-selection and an integral-cohesion. It follows that *age* **of** *algol* may appear as a destination (8.3.1.1.b) in an assignation, but *age* **of** *pl1* may not. Similarly, *algol* is a reference-to-[language]-base, but *pl1* is a [language]-base and no assignment may be made to *pl1*. (Here, [language] stands for structured-with-integral-field-[age]-and-reference-to-[language]-field-[father] and [age] stands for letter-a-letter-g-letter-e, etc.) The selection *father* **of** *pl1*, however, is a reference-to-[language]-selection and thus a reference-to-[language]-cohesion whose value is the name possessed by *algol*. It follows that the identity-relation *father* **of** *pl1* :=: *algol* possesses the value *true*. If *father* **of** *pl1* is used as a destination in an assignation, then there is no change in the name which is a field of the structured value possessed by *pl1*, but there may well be a change in the value of mode [language] referred to by that name. By similar reasoning and because the operators **re** and **im** possess routines (10.2.7.b,c) which deliver values whose mode is 'real' and not 'reference to real', *re* **of** *z* := **im** *w* is an assignation, but **re** *z* := **im** *w* is not.}

8.6. Bases

{Bases are mode-identifiers, e.g., *x*, denotations, e.g., *3.14*, slices, e.g., *x1 [i]* and calls, e.g., *sin (x)*. Bases are, generally, elaborated first. They may be subscripted, parametrized and selected from and are often used as operands. Moreover, certain void-bases are void-cast-packs, e.g., *(: x := x + 1)*, which may be used, e.g., as procedured-coercends; it is essential that they begin with an open-symbol and end with a close-symbol for, otherwise, the parsing of, e.g., *a* := :*b*, which might, in practice, be indistinguishable from *a* :=: *b*, would depend on the modes of *a* and *b*.}

8.6.0.1. Syntax

a) MODE base{81d,820d,e,f,g,821a,b,822a,b,c,823a,824b,d,825a,b,c,d,826a, 828b} : MODE mode identifier{41b} ; MODE denotation{510b,511a,512a, 513a,514a,52b,c,53b,54b,55a,—} ; MODE slice{861a} ; MODE call {862a}.

b) void base{81d,820d,e,f,g,823a} : void call{862a} ; void cast{834a} pack.

{Examples:

a) *x* ; *3.14* ; *x2 [i, j]* ; *sin (x)* ;

b) *lock (stand in)* ; *(: x := 3.14)* }

8.6.0.2. Semantics

a) A mode-identifier is elaborated by making a copy of the instance of the value, if any, possessed by the defining occurrence identified by it {4.1.2, 7.4.2.Step 7}; its value is the copy.

b) The elaboration of a void-cast-pack is that of its void-cast.

8.6.1. Slices

{Slices are obtained by subscripting, e.g., *x1 [i]* or by trimming, e.g., *x1 [2 : n]*, or by a mixture of both, e.g., *x2 [j : n, j]* or *x2 [, k]*. Subscripting and trimming may be done only to primaries, e.g., *x1* and *x2* or *(p | x1 | y1)*. The value of a slice may be either one element of the value of its primary, e.g., *x1 [i]* is a real number from the row of real numbers *x1*, or a subset of the elements, e.g., *x2 [i]* is the *i*-th row of the matrix *x2* and *x2 [, k]* is the *k*-th column.}

8.6.1.1. Syntax

a) REFETY ROWSETY ROWWSETY NONROW slice{860a} :
weak REFETY ROWS ROWWSETY NONROW primary{81d}, sub symbol {31e}, ROWS leaving ROWSETY indexer{b,c,d,e,—}, bus symbol{31e}.

b) row of ROWS leaving row of ROWSETY indexer{a,b} :
trimmer{f}, comma symbol{31e},
ROWS leaving ROWSETY indexer{b,c,d,e,—} ;
subscript{i}, comma symbol{31e},
ROWS leaving row of ROWSETY indexer{b,d,—}.

c) row of ROWS leaving EMPTY indexer{a,b,c} :
subscript{i}, comma symbol{31e}, ROWS leaving EMPTY indexer{c,e}.

d) row of leaving row of indexer{a,b} : trimmer{f}.

e) row of leaving EMPTY indexer{a,b,c} : subscript{i}.

f) trimmer{b,d} : strict lower bound{71u} option, up to symbol{31e},
strict upper bound{71u} option, new lower bound part{g} option.

g) new lower bound part{f} : at symbol{31e}, new lower bound{h}.

h) new lower bound{g} : strong integral tertiary{81b}.

i) subscript{b,c,e} : strong integral tertiary{81b}.

j)* trimscript : trimmer{f} ; subscript{i}.

k)* indexer : ROWS leaving ROWSETY indexer{b,c,d,e}.

l)* boundscript : strict LOWPER bound{71u} ; new lower bound{h} ;
subscript{i}.

{Examples:

a) *x1* [*i*] ; *x2* [*i*, *j*] ; *x2* [*i*] ; *x1* [*2* : *n*] ;

b) *2* : *n*, *j* ; *1*, *2* : *n* ;

c) *i*, *j* ;

d) *2* : *n* ;

e) *i* ;

f) *2* : *n* ; *2* : *n* @ *0* ;

g) @ *0* ;

h) *0* ;

i) *i*}

{In rule a, 'ROWS' reflects the number of trimscripts in the slice, 'ROWSETY' the number of these which are trimmers and 'ROWWSETY' the number of 'row of' not involved in the indexer. In the slices *x2* [*i*, *j*], *x2* [*i*, *2* : *n*], *x2* [*i*], these numbers are (2, 0, 0), (2, 1, 0) and (1, 0, 1) respectively. Because of rules f and 7.1.1.u, *2* : *3* @ *0*, *2* : *n*, *2* :, : *5* and : @ *0* are trimmers.}

8.6.1.2. Semantics

A slice is elaborated in the following steps:

Step 1: Its primary, and all its constituent boundscripts not contained in its primary, are elaborated collaterally {6.2.2.a}; if the value of the primary is ***nil***, then the further elaboration is undefined; otherwise, Step 2 is taken;

Step 2: The multiple value which is, or is referred to by, the value of the primary, is considered, a copy is made of its descriptor, and all the states {2.2.3.3.b} in the copy are set to *1*;

Step 3: The trimscript following the sub-symbol is considered and a pointer, *i*, is set to *1*;

Step 4: If the considered trimscript is an up-to-symbol, then Step 6 is taken; otherwise, if it is a trimmer, then Step 5 is taken; otherwise, letting k stand for its value, if $l_i \leq k \leq u_i$, then the offset in the copy is increased by $(k - l_i) \times d_i$, the i-th quintuple is "marked", and Step 6 is taken; otherwise, the further elaboration is undefined;

Step 5: The values l, u and l' are determined from the considered trimscript as follows:

if the considered trimscript contains a strict-lower-bound (strict-upper-bound), then l (u) is its value; otherwise, l (u) is l_i (u_i); if it contains a new-lower-bound, then l' is its value; otherwise, l' is 1; if now $l_i \leq l$ and $u \leq u_i$, then the offset in the copy is increased by $(l - l_i) \times d_i$, and then l_i is replaced by l' and u_i by $(l' - l) + u$; otherwise, the further elaboration is undefined;

Step 6: If the considered trimscript is followed by a comma-symbol, then the trimscript following that comma-symbol is considered instead, i is increased by 1, and Step 4 is taken; otherwise, all quintuples in the copy which were marked by Step 4 are removed, and Step 7 is taken;

Step 7: If the copy now contains at least one quintuple, then the multiple value composed of the copy and those elements of the considered value which it describes and whose mode is obtained by deleting the initial 'reference to', if any, from the mode enveloped by the original of the slice, is considered instead; otherwise, the element of the considered value selected by {the index equal to} the offset in the copy is considered instead;

Step 8: If the value of the primary is a name, then the value of the slice is a new instance of the name which refers to the considered value, and, otherwise, is a new instance of the considered value itself.

{A trimmer restricts the possible values of a subscript and changes its notation: first, the value of the subscript is restricted to run from the value of the strict-lower-bound to the value of the strict-upper-bound, both given in the old notation; next, all restricted values of that subscript are changed by adding the same amount to each of them, such that the lowest value then equals the value of the new-lower-bound. Thus, the assignations *y1[1 : n − 1] := x1[2 : n]; y1[n] := x1[1]; x1 := y1* effect a cyclic permutation of the elements of *x1*.}

8.6.2. Calls

{Calls are obtained by parametrizing, e.g., *sin (x + 1)*. Parametrizing may be done only to primaries, e.g., *sin* and *cos* or *(p | sin | cos)*. The completed elaboration of a call may or may not deliver a value.}

8.6.2.1. Syntax

a) MOID call{860a,b} : firm procedure with PARAMETERS MOID primary{81d}, actual PARAMETERS{54c,74b} pack.

{Example:

a) *sin (x)* }

8.6.2.2. Semantics

A call is elaborated in the following steps:

Step 1: Its primary is elaborated;

Step 2: The call is replaced by a closed-clause which is a copy of {the routine which is} the value obtained in Step 1;

Step 3: The constituent serial-clause of the closed-clause is protected {6.0.2.d};

Step 4: The skip-symbols {5.4.2.Step 2} following the equals-symbols following the copied formal-parameters are replaced in the textual order by copies of the constituent actual-parameters of the call taken in the same order;

Step 5: The closed-clause as modified in Steps 3 and 4 is replaced by a closed-clause consisting of the same sequence of symbols; the elaboration of this closed-clause is initiated; its value, if any, is then that of the call and if this elaboration is completed or terminated, then this closed-clause is replaced by the call before the elaboration of a successor is initiated.

{The call *samelson (m, (***int** *j)* **real** : *x1 [j])* in the reach of the declaration

```
proc samelson = (int n, proc (int) real f) real :
  begin long real s := long 0; for i to n do s +:= leng f (i) ↑ 2;
        short long sqrt (s)
   end
```

is elaborated by replacing it (Step 2) by the closed-clause

```
(int n = ∼, proc (int) real f = ∼; real :
  begin long real s := long 0; for i to n do s +:= leng f (i) ↑ 2;
        short long sqrt (s)
  end ).
```

Supposing that *n*, *s*, *f* and *i* do not occur elsewhere in the program, this closed-clause is protected (Step 3) without further alteration. The actual-parameters are now inserted (Step 4), yielding the closed-clause

```
(int n = m, proc (int) real f = (int j) real : x1 [j]; real :
  begin long real s := long 0; for i to n do s +:= leng f (i) ↑ 2;
        short long sqrt (s)
  end ),
```

and this closed-clause is elaborated (Step 5). Note that, for the duration of this elaboration, *n* possesses the same integer as that referred to by the name possessed by *m*, and *f* possesses the same routine as that possessed by the routine-denotation *((***int** *j)* **real** : *x1 [j])*. During the elaboration of this and its inner nested closed-clauses (9.3), the elaboration of *f (i)* itself involves the elaboration of the closed-clause *(***int** *j = i;* **real** : *x1 [j])*, and, within this inner closed-clause, the first occurrence of *j* possesses the same integer as that possessed by *i*.}

9. Extensions

a) An extension is the insertion of a comment between two symbols or the replacement of a certain sequence of symbols, possibly satisfying certain restrictions, by another sequence of symbols, as indicated in Sections 9.1 up to 9.4.

b) No extension may be performed within a comment {3.0.9.b}, character-denotation {5.1.4.1.a}, or row-of-character-denotation {5.3.1.b}.

c) Some extensions are given in the representation language, except that
A, *B* and *C* stand for strong-unitary-integral-clauses {8.1.1.a},
D for a strong-serial-boolean-clause {6.1.1.a},
E for a strong-unitary-void-clause {8.1.1.a},
F and *G* for unitary-clauses {8.1.1.a},
H for two or more unitary-clauses {8.1.1.a} separated by comma-symbols {3.1.1.c},
I, *J*, *K* and *L* for mode-identifiers {4.1.1.b},
M for a label-identifier {4.1.1.b},
N for a local-symbol {3.1.1.d} or heap-symbol{3.1.1.d}-option,
O for a conformity-relator {8.3.2.1.b},
P for an indication {4.2.1.a},
Q for a virtual-plan {7.1.1.x,aa},
R for a routine-denotation {5.4.1.a},
S for the standard-prelude {2.1.b, 10} if the extension is performed outside the standard-prelude and, otherwise, for the empty sequence of symbols,
T for a condition followed by a choice-clause {6.4.1.b,c,d},
U for a declarer {7.1.1.a},
V for a virtual-declarer {7.1.1.b} or for a formal-declarer {7.1.1.b} all of whose constituent formal-lower-bounds and formal-upper-bounds are either-symbols,
W, *X* and *Y* for tertiaries {8.1.1.b},
Z for two or more tertiaries {8.1.1.b} separated by comma-symbols {3.1.1.e},
Γ for a comma-symbol {3.1.1.e}, go-on-symbol {3.1.1.f} or becomes-symbol {3.1.1.c},
Σ for a serial-clause {6.1.1.a},
Φ for a VICTAL-ROWS-rower {7.1.1.q,r} or indexer {8.6.1.1.k}, where "VICTAL" ("ROWS") stands for any terminal production of the metanotion 'VICTAL' ('ROWS'),
Δ for an open-symbol {3.1.1.e}, and
∇ for a close-symbol {3.1.1.e}.

d) Each representation of a symbol appearing in sections 9.1 up to 9.4 may be replaced by any other representation, if any, of the same symbol.

9.1. Comments

{*A source of innocent merriment.*
Mikado, *W. S. Gilbert.*}

A comment {3.0.9.b} may be inserted between any two symbols {but see 9.b}.
{e.g., *(m>n | m | n)* may be replaced by
(m>n | m # the larger of the two # | n).}

9.2. Contractions

a) **ref** *V I* = **loc** *U* Γ and **ref** *V I* = **heap** *U* Γ where **ref** *V I* is the formal-parameter of an identity-declaration {7.4.1.a} which is not followed by a comma-symbol {3.1.1.e} followed by a mode-identifier {4.1.1.b} and where U and V specify the same mode {7.1.2.a} may be replaced by $U\,I\,\Gamma$ and **heap** $U\,I\,\Gamma$ respectively.

{e.g., **ref real** *x* = **loc real**; may be replaced by **real** *x*; , **ref bool** *p* = **loc bool** := **true** may be replaced by **bool** *p* := **true**, and **ref real** *t* = **heap real**; may be replaced by **heap real** *t*.}

b) **mode** P = **struct** may be replaced by **struct** P = and **mode** P = **union** by **union** P = .

{e.g., **mode compl** = **struct** (**real** *re, im*) (see also 9.2.c) may be replaced by **struct compl** = (**real** *re, im*).}

c) If a given mode-declaration {7.2.1.a} (priority-declaration {7.3.1.a}, identity-declaration {7.4.1.a}, operation-declaration {7.5.1.a}, formal-parameter {5.4.1.e}, field-declarator {7.1.1.g}) and another one following a comma-symbol {3.1.1.e} following the given one both begin with a mode-symbol, structure-symbol, union-of-symbol, priority-symbol, operation-symbol {all 3.1.1.d}, or one same terminal production of 'VICTAL MODE declarer' {7.1.1.b} or of 'MODE global generator' {8.5.1.1.c} where "MODE" ("VICTAL") stands for any terminal production of the metanotion 'MODE' ('VICTAL'), then the second of these occurrences may be omitted.

{e.g., **real** *x*, **real** *y* := *1.2* may be replaced by **real** *x, y* := *1.2*, but **real** *x*, **real** *y* = *1.2* may not be replaced by **real** *x, y* = *1.2*, since the first occurrence of **real** is an actual-declarer whereas the second is a formal-declarer. Note also that **mode b** = **bool**, **mode r** = **real** may be replaced by **mode b** = **bool**, **r** = **real**, etc.}

d) If an actual-parameter {7.4.1.b} (source {8.3.1.1.c}) is a routine-denotation {5.4.1.a} or a void-cast-pack {8.3.4.1.a} (is a routine-denotation), then its first open-symbol and last close-symbol {both 3.1.1.e} may simultaneously be omitted.

{e.g., **op** + = ((**int** *a*) **int** : *a*) may be replaced by **op** + = (**int** *a*) **int** : *a*.}

e) If the original {1.1.6.c} of Q and the original of R envelop {1.1.6.j} the same mode, then the unitary-phrase **proc** Q I = R (**op** Q P = R, N **proc** Q := R) may be replaced by **proc** I = R (by **op** P = R, by N **proc** := R).

{e.g., **proc** (**ref int**) *incr* = (**ref int** *i*) : *i* +:= *1* may be replaced by **proc** *incr* = (**ref int** *i*) : *i* +:= *1*, **op** (**ref int**) **int decr** = (**ref int** *i*) **int** : *i* −:= *1* may be replaced by **op decr** = (**ref int** *i*) **int** : *i* −:= *1*, and the actual-parameter of the identity-declaration **ref proc** (**real**) **int** *p* = **loc proc** (**real**) **int** := ((**real** *x*) **int** : **round** *x*) may be replaced by **loc proc** := ((**real** *x*) **int** : **round** *x*), whereupon application of 9.2.a,d may yield the identity-declaration **proc** *p* := (**real** *x*) **int** : **round** *x*.}

f) [:] may be replaced by [ ], [:, may be replaced by [, , ,:, may be replaced by ,, , ,:] may be replaced by ,] , [: @ may be replaced by [@ , and ,: @ may be replaced by ,@ .

{e.g., [:] **real** may be replaced by [ ] **real**.}

g) [Φ] may be replaced by $\Delta\Phi\nabla$ or by $\Delta/\Phi/\nabla$.

{e.g., [*i*] may be replaced by (*i*) or by (/*i*/).}

9.3. Repetitive Statements

a) The strong-unitary-void-clause {8.1.1.a}

begin int $J := A$, **int** $K = B$, $L = C$;
  M: **if** S $(K>0 \mid J \leq L \mid: K<0 \mid J \geq L \mid$ **true**)
    then int $I = J$; $(D \mid E$; $(S\ J +:= K)$; **go to** M)
    fi
  end,

where J, K, L and M do not occur in D, E or S, and where I differs from J and K, may be replaced by

for I **from** A **by** B **to** C **while** D **do** E,

and if, moreover, I does not occur in D or E, then **for** I **from** may be replaced by **from**.

b) The strong-unitary-void-clause {8.1.1.a}

begin int $J := A$, **int** $K = B$;
  M: (**int** $I = J$; $(D \mid E$; $(S\ J +:= K)$; **go to** M))
  end,

where J, K and M do not occur in D, E or S, and where I differs from J and K, may be replaced by

for I **from** A **by** B **while** D **do** E,

and if, moreover, I does not occur in D or E, then **for** I **from** may be replaced by **from**.

c) **from** *1* **by** may be replaced by **by**, **by** *1* **to** by **to**, **by** *1* **while** by **while**, and **while true do** by **do**.

{e.g., **for** i **from** *1* **by** *1* **to** n **while true do** $x +:= x1[i]$ may be replaced by **for** i **to** n **do** $x +:= x1[i]$. Note that **to** *0* **do** E and **while false do** E do not cause E to be elaborated at all, whereas **do** E causes E to be elaborated repeatedly until the elaboration is terminated, interrupted or halted.}

9.4. Contracted Conditional Clauses

{*The flowers that bloom in the spring,*
Tra la,
Have nothing to do with the case.
Mikado, *W. S. Gilbert.*}

a) **else skip fi** may be replaced by **fi**.

{e.g., **if** $x<0$ **then** $x := 0$ **else skip fi** may be replaced by **if** $x<0$ **then** $x := 0$ **fi**.}

b) **else if** T **fi fi** may be replaced by **elsf** T **fi** and
then if T **fi fi** may be replaced by **thef** T **fi**.

{e.g., **if** p **then** *princeton* **else if** q **then** *grenoble* **else** *zandvoort* **fi fi** may be replaced by **if** p **then** *princeton* **elsf** q **then** *grenoble* **else** *zandvoort* **fi** or by (p | *princeton* |: q | *grenoble* | *zandvoort*). Many more examples are to be found in 10.5.}

c) *(***int** $I = A$; **if** $S\ I = 1$ **then** F **elsf** $S\ I = 2$ **then** G **else** Σ **fi***)*, where I does not occur in F, G, S or Σ, may be replaced by **case** A **in** F, G **out** Σ **esac** {or by $(A \mid F, G \mid \Sigma)$}.

d) *(***int** $I = A$; **if** $S\ I = 1$ **then** F **else case** $(S\ I - 1)$ **in** H **out** Σ **esac fi***)*, where I does not occur in F, H, S or Σ, may be replaced by **case** A **in** F, H **out** Σ **esac** {or by $(A \mid F, H \mid \Sigma)$}.

{Examples of the use of such "case clauses" are given in 11.11.v,am and 11.12.}

e) *(***int** I, **sema** $L = (S\ /\ 1)$; $V\ K = W$;
    par $((X\ O\ K \mid S \downarrow L;\ I := 1;\ M),$
        $(Y\ O\ K \mid S \downarrow L;\ I := 2;\ M));\ 0.\ M\!:\ I)$,

where W is the same as some terminal production of 'MODE tertiary' in which "MODE" stands for the mode specified by V, and where I, K, L and M do not occur in W, X and Y, may be replaced by $[\star X, Y\ O\ W \star]$.

f) *(***int** I, J; **sema** $L = (S\ /\ 1)$; $V\ K = W$;
    par $((X\ O\ K \mid S \downarrow L;\ I := 1;\ M),$
        $(S\ (J := ([\star Z\ O\ K \star] + 1)) > 1 \mid S \downarrow L;\ I := J;\ M));\ 0.\ M\!:\ I)$,

where W is the same as some terminal production of 'MODE tertiary' in which "MODE" stands for the mode specified by V, and where I, J, K, L and M do not occur in S, W, X or Z, may be replaced by $[\star X, Z\ O\ W \star]$.

g) $([\star Z\ O\ W \star] \mid H \mid \Sigma)$ may be replaced by **case** $Z\ O\ W$ **in** H **out** Σ **esac**.

{Examples of the use of such "conformity case clauses" are given in 11.11.q,ah.}

10. Standard Prelude and Postlude

a) A "standard declaration" is one of the constituent declarations of the standard-prelude {2.1.b} {; it is either an "environment enquiry" supplying information concerning a specific property of the implementation (2.3.c), a "standard priority" or "standard operation", a "standard mathematical constant or function", a "synchronization operation" or a "transput declaration"}.

b) A representation of the standard-prelude is obtained by altering each form in 10.1, 10.2, 10.3, 10.4 and 10.5 in the following steps:

Step 1: Each sequence of symbols between ⟨ and ⟩ in a given form is altered in the following steps:

Step 1.1: If **D** occurs in the given sequence of symbols, then the given sequence is replaced by a chain of a sufficient number of sequences separated by comma-symbols; the first new sequence is a copy of the given sequence in which copy **D** is deleted; the n-th new sequence, $n > 1$, is a copy of the given sequence in which copy **D** is replaced by a sub-symbol followed by $n - 2$ comma-symbols followed by a bus-symbol;

Step 1.2: If, in the given sequence of symbols, as possibly modified in Step 1.1, **L int**, **L real** or **L compl** occurs, then that sequence is replaced by a chain of a sufficient number of sequences separated by comma-symbols, the n-th new sequence being a copy of the given sequence in which copy each occurrence of L (**L**) has been replaced by $(n - 1)$ times *long* (**long**);

Step 2: Each occurrence of ⟨ and ⟩ in a given form, as possibly modified in Step 1, is deleted;

Step 3: If, in a given form, as possibly modified in Steps 1 and 2, **L int** (**L real**, **L compl**, **L bits**, **L bytes**) occurs, then the form is replaced by a sequence of a sufficient number of new forms; the *n*-th new form is a copy of the given form in which copy each occurrence of *L* (**L**, **K**, **S**) is replaced by *(n − 1)* times *long* (**long**, **leng**, **short**);

Step 4: If **P** occurs in a given form, as possibly modified or made in the Steps above, then the form is replaced by four new forms obtained by replacing **P** consistently throughout the form by either − or + or × or /;

Step 5: If **Q** occurs in a given form, as possibly modified or made in the Steps above, then the form is replaced by four new forms obtained by replacing **Q** consistently throughout the form by either −:= or +:= or ×:= or /:=;

Step 6: If **R** occurs in a given form, as possibly modified or made in the Steps above, then the form is replaced by six new forms obtained by replacing **R** consistently throughout the form by either $<$ or $\leq$ or $=$ or $\neq$ or $\geq$ or $>$;

Step 7: If **E** occurs in a given form, as possibly modified or made in the Steps above, then the form is replaced by two new forms obtained by replacing **E** consistently throughout the form by either = or ≠;

Step 8: Each occurrence of *F* in any form, as possibly modified or made in the Steps above, is replaced by a representation of 'letter aleph symbol' {5.5.8};

Step 9: If, in some form, as possibly modified or made in the Steps above, % occurs followed by the representation of an identifier (field-selector, indication), then that occurrence of % is deleted and each occurrence of that representation in any form is replaced by one same representation of an identifier (field-selector, indication) which does not occur elsewhere in the program and Step 9 is taken;

Step 10: If a sequence of representations beginning with and ending with **c** occurs in any form, as possibly modified or made in the Steps above, then this sequence is replaced by a representation of an actual-declarer or closed-clause suggested by the sequence;

Step 11: If, in any form, as possibly modified or made in the Steps above, a representation of a routine-denotation occurs whose elaboration involves the manipulation of real numbers, then this denotation may be replaced by any other denotation whose elaboration has approximately the same effect {; the degree of approximation is left undefined in this Report (see also 2.2.3.1.c)};

Step 12: The standard-prelude is that declaration-prelude-sequence whose representation is the same as the sequence of all the forms, as possibly modified or made in the Steps above.

{The declarations in this Chapter are intended to describe their effect clearly. The effect may very well be obtained by a more efficient method.}

c) A representation of the standard-postlude is given in 10.6.

10.1. Environment Enquiries

a) **int** *int lengths* = **c** *the number of different lengths of integers* **c**;

b) **L int** *L max int* = **c** *the largest L integral value* **c**;

c) **int** *real lengths* =
c *the number of different lengths of real numbers* **c**;

d) **L real** *L max real* = **c** *the largest L real value* **c**;

e) **L real** *L small real* = **c** *the smallest L real value such that both* **L** *1* + *L small real* > **L** *1 and* **L** *1* − *L small real* < **L** *1* **c**;

f) **int** *bits widths* = **c** *the number of different widths of bits* **c**;

g) **int** *L bits width* =
c *the number of elements in L bits; see* **L bits** {10.2.8.a} **c**;

h) **int** *bytes widths* = **c** *the number of different widths of bytes* **c**;

i) **int** *L bytes width* =
c *the number of elements in L bytes; see* **L bytes** {10.2.9.a} **c**;

j) **op abs** = *(***char** *a)* **int**: **c** *the integral equivalent of the character 'a'* **c**;

k) **op repr** = *(***int** *a)* **char**:
c *that character 'x', if it exists, for which* **abs** *x = a* **c**;

l) **char** *null character* = **c** *some character* **c**;

10.2. Standard Priorities and Operations

10.2.0. Standard Priorities

a) **priority** −:= = *1*, +:= = *1*, ×:= = *1*, /:= = *1*, ÷:= = *1*, ÷::= = *1*, +=: = *1*, ∨ = *2*, ∧ = *3*, = = *4*, ≠ = *4*, < = *5*, ≤ = *5*, ≥ = *5*, > = *5*, − = *6*, + = *6*, × = *7*, ÷ = *7*, ÷: = *7*, / = *7*, □ = *7*, ↑ = *8*, ⌊ = *8*, ⌈ = *8*, ⌞ = *8*, ⌜ = *8*, ⊥ = *9*;

10.2.1. Rows and Associated Operations

a) **mode** % **rows** = **c** *an actual-declarer specifying a mode united from* {4.4.3.a} *all modes beginning with 'row of'* **c**;

b) **op** ⌊ = *(***int** *n,* **rows** *a)* **int**: **c** *the lower bound in the n-th quintuple of the descriptor of the value of 'a', if that quintuple exists* **c**;

c) **op** ⌈ = *(***int** *n,* **rows** *a)* **int**: **c** *the upper bound in the n-th quintuple of the descriptor of the value of 'a', if that quintuple exists* **c**;

d) **op** ⌞ = *(***int** *n,* **rows** *a)* **bool**: **c** *true (false) if the lower state in the n-th quintuple of the descriptor of the value of 'a' equals 1 (0), if that quintuple exists* **c**;

e) **op** ⌜ = *(***int** *n,* **rows** *a)* **bool**: **c** *true (false) if the upper state in the n-th quintuple of the descriptor of the value of 'a' equals 1 (0), if that quintuple exists* **c**;

f) **op** ⌊ = *(***rows** *a)* **int**: *1* ⌊ *a*;

g) **op** ⌈ = *(***rows** *a)* **int**: *1* ⌈ *a*;

h) **op** ⌞ = *(***rows** *a)* **bool**: *1* ⌞ *a*;

i) **op** ⌜ = *(***rows** *a)* **bool**: *1* ⌜ *a*;

10.2.2. Operations on Boolean Operands

a) **op** ∨ = (**bool** *a, b*) **bool** : (*a* | **true** | *b*);
b) **op** ∧ = (**bool** *a, b*) **bool** : (*a* | *b* | **false**);
c) **op** ¬ = (**bool** *a*) **bool** : (*a* | **false** | **true**);
d) **op** = = (**bool** *a, b*) **bool** : (*a*∧*b*)∨(¬ *a*∧ ¬*b*);
e) **op** ≠ = (**bool** *a, b*) **bool** : ¬ (*a*=*b*);
f) **op abs** = (**bool** *a*) **int** : (*a* | *1* | *0*);

10.2.3. Operations on Integral Operands

a) **op** < = (**L int** *a, b*) **bool** : **c** *true if the value of 'a' is smaller than* {2.2.3.1.c} *that of 'b' and false otherwise* **c**;
b) **op** ≤ = (**L int** *a, b*) **bool** : ¬ (*b*<*a*);
c) **op** = = (**L int** *a, b*) **bool** : *a*≤*b*∧*b*≤*a*;
d) **op** ≠ = (**L int** *a, b*) **bool** : ¬ (*a*=*b*);
e) **op** ≥ = (**L int** *a, b*) **bool** : *b*≤*a*;
f) **op** > = (**L int** *a, b*) **bool** : *b*<*a*;
g) **op** − = (**L int** *a, b*) **L int** : **c** *the value of 'a' minus* {2.2.3.1.c} *that of 'b'* **c**;
h) **op** − = (**L int** *a*) **L int** : **L** *0*−*a*;
i) **op** + = (**L int** *a, b*) **L int** : *a*− −*b*;
j) **op** + = (**L int** *a*) **L int** : *a*;
k) **op abs** = (**L int** *a*) **L int** : (*a*<**L** *0* | −*a* | *a*);
l) **op** × = (**L int** *a, b*) **L int** : (**L int** *s* := **L** *0*, *i* := **abs** *b*;
while *i*≥**L** *1* **do** (*s* := *s*+*a*; *i* := *i*−**L** *1*); (*b*<**L** *0* | −*s* | *s*));
m) **op** ÷ = (**L int** *a, b*) **L int** : (*b*≠**L** *0* | **L int** *q* := **L** *0*, *r* := **abs** *a*;
while (*r* := *r*−**abs** *b*)≥**L** *0* **do** *q* := *q*+**L** *1*;
(*a*<**L** *0*∧*b*≥**L** *0*∨*a*≥**L** *0*∧*b*<**L** *0* | −*q* | *q*));
n) **op** ÷: = (**L int** *a, b*) **L int** : (**int** *r*=*a*−*a*÷*b*×*b*; (*r*<*0* | *r*+**abs** *b* | *r*));
o) **op** / = (**L int** *a, b*) **L real** : (**L real** : *a*) / (**L real** : *b*);
p) **op** ↑ = (**L int** *a*, **int** *b*) **L int** :
(*b*≥*0* | **L int** *p* := **L** *1*; **to** *b* **do** *p* := *p*×*a*; *p*);
q) **op leng** = (**L int** *a*) **long L int** : **c** *the long L integral value equivalent to* {2.2.3.1.d} *the value of 'a'* **c**;
r) **op short** = (**long L int** *a*) **L int** : **c** *the L integral value, if it exists, equivalent to* {2.2.3.1.d} *the value of 'a'* **c**;
s) **op odd** = (**L int** *a*) **bool** : **abs** *a* ÷: **L** *2* = **L** *1*;
t) **op sign** = (**L int** *a*) **int** : (*a*>**L** *0* | *1* |: *a*<**L** *0* | −*1* | *0*);
u) **op** ⊥ = (**L int** *a, b*) **L compl** : (*a, b*);

10.2.4. Operations on Real Operands

a) **op** < = (**L real** *a, b*) **bool** : **c** *true if the value of 'a' is smaller than* {2.2.3.1.c} *that of 'b' and false otherwise* **c**;
b) **op** ≤ = (**L real** *a, b*) **bool** : ¬ (*b*<*a*);
c) **op** = = (**L real** *a, b*) **bool** : *a*≤*b*∧*b*≤*a*;
d) **op** ≠ = (**L real** *a, b*) **bool** : ¬ (*a*=*b*);
e) **op** ≥ = (**L real** *a, b*) **bool** : *b*≤*a*;
f) **op** > = (**L real** *a, b*) **bool** : *b*<*a*;

g) **op** − = *(***L real** *a, b)* **L real** : **c** *the value of 'a' minus* {2.2.3.1.c} *that of 'b'* **c**;
h) **op** − = *(***L real** *a)* **L real** : **L** *0* − *a;*
i) **op** + = *(***L real** *a, b)* **L real** : *a* − −*b;*
j) **op** + = *(***L real** *a)* **L real** : *a;*
k) **op abs** = *(***L real** *a)* **L real** : *(a* < **L** *0* | −*a* | *a);*
l) **op** × = *(***L real** *a, b)* **L real** : **c** *the value of 'a' times* {2.2.3.1.c} *that of 'b'* **c**;
m) **op** / = *(***L real** *a, b)* **L real** : **c** *the value of 'a' divided by* {2.2.3.1.c} *that of 'b'* **c**;
n) **op leng** = *(***L real** *a)* **long L real** : **c** *the long L real value equivalent to* {2.2.3.1.d} *the value of 'a'* **c**;
o) **op short** = *(***long L real** *a)* **L real** : **c** *if* **abs** *a* ≤ **leng** *L max real, then a L real value 'v' such that, for any L real value 'w',* **abs** *(***leng** *v* − *a)* ≤ **abs** *(***leng** *w* − *a)* **c**;
p) **op round** = *(***L real** *a)* **L int** : **c** *a L integral value, if one exists, equivalent to* {2.2.3.1.d} *a L real value differing by not more than one-half from the value of 'a'* **c**;
q) **op sign** = *(***L real** *a)* **int** : *(a* > **L** *0* | *1* |: *a* < **L** *0* | −*1* | *0);*
r) **op entier** = *(***L real** *a)* **L int** : *(***L int** *j* := **L** *0;* **while** *j* < *a* **do** *j* := *j* + **L** *1;* **while** *j* > *a* **do** *j* := *j* − **L** *1; j);*
s) **op** ⊥ = *(***L real** *a, b)* **L compl** : *(a, b);*

10.2.5. Operations on Arithmetic Operands

a) **op P** = *(***L real** *a,* **L int** *b)* **L real** : *a* **P** *(***L real** : *b);*
b) **op P** = *(***L int** *a,* **L real** *b)* **L real** : *(***L real** : *a)* **P** *b;*
c) **op R** = *(***L real** *a,* **L int** *b)* **bool** : *a* **R** *(***L real** : *b);*
d) **op R** = *(***L int** *a,* **L real** *b)* **bool** : *(***L real** : *a)* **R** *b;*
e) **op** ⊥ = *(***L real** *a,* **L int** *b)* **L compl** : *(a, b);*
f) **op** ⊥ = *(***L int** *a,* **L real** *b)* **L compl** : *(a, b);*
g) **op** ↑ = *(***L real** *a,* **int** *b)* **L real** : *(***L real** *p* := **L** *1;* **to abs** *b* **do** *p* := *p* × *a; (b* ≥ *0* | *p* | **L** *1* / *p));*

10.2.6. Operations on Character Operands

a) **op R** = *(***char** *a, b)* **bool** : **abs** *a* **R abs** *b;* {10.1.j}
b) **op** + = *(***char** *a, b)* **string** : *(a, b);*

10.2.7. Complex Structures and Associated Operations

a) **struct L compl** = *(***L real** *re, im);*
b) **op re** = *(***L compl** *a)* **L real** : *re* **of** *a;*
c) **op im** = *(***L compl** *a)* **L real** : *im* **of** *a;*
d) **op abs** = *(***L compl** *a)* **L real** : *L sqrt (***re** *a* ↑ *2* + **im** *a* ↑ *2);*
e) **op conj** = *(***L compl** *a)* **L compl** : **re** *a* ⊥ − **im** *a;*
f) **op** = = *(***L compl** *a, b)* **bool** : **re** *a* = **re** *b* ∧ **im** *a* = **im** *b;*
g) **op** ≠ = *(***L compl** *a, b)* **bool** : ¬ *(a* = *b);*
h) **op** − = *(***L compl** *a, b)* **L compl** : *(***re** *a* − **re** *b)* ⊥ *(***im** *a* − **im** *b);*
i) **op** − = *(***L compl** *a)* **L compl** : − **re** *a* ⊥ − **im** *a;*
j) **op** + = *(***L compl** *a, b)* **L compl** : *(***re** *a* + **re** *b)* ⊥ *(***im** *a* + **im** *b);*

k) **op** + = (**L compl** *a*) **L compl** : *a*;
l) **op** × = (**L compl** *a*, *b*) **L compl** :
(**re** *a* × **re** *b* − **im** *a* × **im** *b*) ⊥ (**re** *a* × **im** *b* + **im** *a* × **re** *b*);
m) **op** / = (**L compl** *a*, *b*) **L compl** : (**L real** *d* = **re** (*b* × **conj** *b*);
L compl *n* = *a* × **conj** *b*; (**re** *n* / *d*) ⊥ (**im** *n* / *d*));
n) **op leng** = (**L compl** *a*) **long L compl** : **leng re** *a* ⊥ **leng im** *a*;
o) **op short** = (**long L compl** *a*) **L compl** : **short re** *a* ⊥ **short im** *a*;
p) **op P** = (**L compl** *a*, **L int** *b*) **L compl** : *a* **P** (**L compl** : *b*);
q) **op P** = (**L compl** *a*, **L real** *b*) **L compl** : *a* **P** (**L compl** : *b*);
r) **op P** = (**L int** *a*, **L compl** *b*) **L compl** : (**L compl** : *a*) **P** *b*;
s) **op P** = (**L real** *a*, **L compl** *b*) **L compl** : (**L compl** : *a*) **P** *b*;
t) **op** ↑ = (**L compl** *a*, **int** *b*) **L compl** : (**L compl** *p* := **L** *1*;
to abs *b* **do** *p* := *p* × *a*; (*b* ≥ *0* | *p* | **L** *1* / *p*));
u) **op E** = (**L compl** *a*, **L int** *b*) **bool** : *a* **E** (**L compl** : *b*);
v) **op E** = (**L compl** *a*, **L real** *b*) **bool** : *a* **E** (**L compl** : *b*);
w) **op E** = (**L int** *a*, **L compl** *b*) **bool** : *b* **E** *a*;
x) **op E** = (**L real** *a*, **L compl** *b*) **bool** : *b* **E** *a*;

10.2.8. Bits Structures and Associated Operations

a) **struct L bits** = ([*1* : *L bits width*] **bool** *L F*); {See 10.1.g}
{The field-selector is hidden from the user in order that he may not break open the structure; in particular, he may not subscript the field.}
b) **op** = = (**L bits** *a*, *b*) **bool** : (**for** *i* **to** *L bits width* **do**
((*L F* **of** *a*) [*i*] ≠ (*L F* **of** *b*) [*i*] | *l*); **true**. *l*: **false**);
c) **op** ≠ = (**L bits** *a*, *b*) **bool** : ¬ (*a* = *b*);
d) **op** ∨ = (**L bits** *a*, *b*) **L bits** : (**L bits** *c*; **for** *i* **to** *L bits width* **do**
(*L F* **of** *c*) [*i*] := (*L F* **of** *a*) [*i*] ∨ (*L F* **of** *b*) [*i*]; *c*);
e) **op** ∧ = (**L bits** *a*, *b*) **L bits** : (**L bits** *c*; **for** *i* **to** *L bits width* **do**
(*L F* **of** *c*) [*i*] := (*L F* **of** *a*) [*i*] ∧ (*L F* **of** *b*) [*i*]; *c*);
f) **op** ≤ = (**L bits** *a*, *b*) **bool** : (*a* ∨ *b*) = *b*;
g) **op** ≥ = (**L bits** *a*, *b*) **bool** : *b* ≤ *a*;
h) **op** ↑ = (**L bits** *a*, **int** *b*) **L bits** : **if abs** *b* ≤ *L bits width* **then L bits** *c* := *a*;
to abs *b* **do** (*b* > *0* | **for** *i* **from** *2* **to** *L bits width* **do** (*L F* **of** *c*) [*i* − *1*] :=
(*L F* **of** *c*) [*i*]; (*L F* **of** *c*) [*L bits width*] := **false** | **for** *i* **from** *L bits width*
by − *1* **to** *2* **do** (*L F* **of** *c*) [*i*] := (*L F* **of** *c*) [*i* − *1*]; (*L F* **of** *c*) [*1*] :=
false); *c* **fi**;
i) **op abs** = (**L bits** *a*) **L int** : (**L int** *c* := **L** *0*; **for** *i* **to** *L bits width* **do**
c := **L** *2* × *c* + **K abs** (*L F* **of** *a*) [*i*]; *c*);
j) **op bin** = (**L int** *a*) **L bits** : **if** *a* ≥ **L** *0* **then L int** *b* := *a*; **L bits** *c*;
for *i* **from** *L bits width* **by** − *1* **to** *1* **do**
((*L F* **of** *c*) [*i*] := **odd** *b*; *b* := *b* ÷ **L** *2*); *c* **fi**;
k) **op** ▯ = (**int** *a*, **L bits** *b*) **bool** : (*L F* **of** *b*) [*a*];
l) **op L btb** = ([*1* :] **bool** *a*) **L bits** : (**int** *n* = ⌈ *a*; (*n* ≤ *L bits width* |
L bits *c*; **for** *i* **to** *L bits width* **do** (*L F* **of** *c*) [*i*] :=
(*i* ≤ *L bits width* − *n* | **false** | *a* [*i* − *L bits width* + *n*]); *c*));
m) **op** ¬ = (**L bits** *a*) **L bits** : (**L bits** *c*; **for** *i* **to** *L bits width* **do**
(*L F* **of** *c*) [*i*] := ¬ (*L F* **of** *a*) [*i*]; *c*);

10.2.9. Bytes and Associated Operations

a) **struct L bytes** = ([1 : L bytes width] **char** L F); {See 10.2.8.a and 10.1.i}
b) **op R** = (**L bytes** a, b) **bool** : (**string** : a) **R** (**string** : b);
c) **op □** = (**int** a, **L bytes** b) **char** : (L F **of** b) [a];
d) **op L ctb** = (**string** a) **L bytes** : (**int** n = ⌈ a; (n ≤ L bytes width |
  L bytes c; **for** i **to** L bytes width **do**
  (L F **of** c) [i] := (i ≤ n | a [i] | null character); c));

10.2.10. Strings and Associated Operations

a) **mode string** = [1 : 0 **flex**] **char**;
b) **op** < = (**string** a, b) **bool** : (**int** m = ⌈ a, n = ⌈ b;
  int p = (m < n | m | n), **int** i := 1; **bool** c;
  (p < 1 | n ≥ 1 | e: (c := a [i] = b [i] |: (i := i + 1) ≤ p | e);
  (c | m < n | a [i] < b [i])));
c) **op** ≤ = (**string** a, b) **bool** : ¬ (b < a);
d) **op** = = (**string** a, b) **bool** : a ≤ b ∧ b ≤ a;
e) **op** ≠ = (**string** a, b) **bool** : ¬ (a = b);
f) **op** ≥ = (**string** a, b) **bool** : b ≤ a;
g) **op** > = (**string** a, b) **bool** : b < a;
h) **op R** = (**string** a, **char** b) **bool** : a **R** (**string** : b);
i) **op R** = (**char** a, **string** b) **bool** : (**string** : a) **R** b;
j) **op** + = (**string** a, b) **string** : (**int** m = ⌈ a, n = ⌈ b; [1 : m + n] **char** c;
  c [1 : m] := a; c [m + 1 : m + n] := b; c);
k) **op** + = (**string** a, **char** b) **string** : a + (**string** : b);
l) **op** + = (**char** a, **string** b) **string** : (**string** : a) + b;

{The operations defined in b, h and i imply that if **abs** "a" < **abs** "b", then "" < "a"; "a" < "b"; "aa" < "ab"; "aa" < "ba"; "ab" < "b".}

10.2.11. Operations Combined with Assignations

a) **op** −:= = (**ref L int** a, **L int** b) **ref L int** : a := a − b;
b) **op** −:= = (**ref L real** a, **L real** b) **ref L real** : a := a − b;
c) **op** −:= = (**ref L compl** a, **L compl** b) **ref L compl** : a := a − b;
d) **op** +:= = (**ref L int** a, **L int** b) **ref L int** : a := a + b;
e) **op** +:= = (**ref L real** a, **L real** b) **ref L real** : a := a + b;
f) **op** +:= = (**ref L compl** a, **L compl** b) **ref L compl** : a := a + b;
g) **op** ×:= = (**ref L int** a, **L int** b) **ref L int** : a := a × b;
h) **op** ×:= = (**ref L real** a, **L real** b) **ref L real** : a := a × b;
i) **op** ×:= = (**ref L compl** a, **L compl** b) **ref L compl** : a := a × b;
j) **op** ÷:= = (**ref L int** a, **L int** b) **ref L int** : a := a ÷ b;
k) **op** ÷::= = (**ref L int** a, **L int** b) **ref L int** : a := a ÷: b;
l) **op** /:= = (**ref L real** a, **L real** b) **ref L real** : a := a / b;
m) **op** /:= = (**ref L compl** a, **L compl** b) **ref L compl** : a := a / b;
n) **op Q** = (**ref L real** a, **L int** b) **ref L real** : a **Q** (**L real** : b);
o) **op Q** = (**ref L compl** a, **L int** b) **ref L compl** : a **Q** (**L compl** : b);
p) **op Q** = (**ref L compl** a, **L real** b) **ref L compl** : a **Q** (**L compl** : b);
q) **op** +:= = (**ref string** a, **string** b) **ref string** : a := a + b;

r) **op** +=: = (**string** *a*, **ref string** *b*) **ref string** : *b* := *a* + *b*;
s) **op** +:= = (**ref string** *a*, **char** *b*) **ref string** : *a* := *a* + *b*;
t) **op** +=: = (**char** *a*, **ref string** *b*) **ref string** : *b* := *a* + *b*;

10.3. Standard Mathematical Constants and Functions

a) **L real** *L pi* = **c** *a L real value close to π; see Math. of Comp. v. 16, 1962, pp. 80—99* **c**;
b) **proc** *L sqrt* = (**L real** *x*) **L real** : **c** *if x* ≥ **L** *0, a L real value close to the square root of 'x'* **c**;
c) **proc** *L exp* = (**L real** *x*) **L real** : **c** *a L real value, if one exists, close to the exponential function of 'x'* **c**;
d) **proc** *L ln* = (**L real** *x*) **L real** : **c** *a L real value, if one exists, close to the natural logarithm of 'x'* **c**;
e) **proc** *L cos* = (**L real** *x*) **L real** : **c** *a L real value close to the cosine of 'x'* **c**;
f) **proc** *L arccos* = (**L real** *x*) **L real** : **c** *if* **abs** *x* ≤ **L** *1, a L real value close to the inverse cosine of 'x',* **L** *0* ≤ *L arccos (x)* ≤ *L pi* **c**;
g) **proc** *L sin* = (**L real** *x*) **L real** : **c** *a L real value close to the sine of 'x'* **c**;
h) **proc** *L arcsin* = (**L real** *x*) **L real** : **c** *if* **abs** *x* ≤ **L** *1, a L real value close to the inverse sine of 'x',* **abs** *L arcsin (x)* ≤ *L pi* / **L** *2* **c**;
i) **proc** *L tan* = (**L real** *x*) **L real** : **c** *a L real value, if one exists, close to the tangent of 'x'* **c**;
j) **proc** *L arctan* = (**L real** *x*) **L real** : **c** *a L real value close to the inverse tangent of 'x',* **abs** *L arctan (x)* ≤ *L pi* / **L** *2* **c**;
k) **proc L real** *L random* = *L last random* := **c** *the next pseudo-random L real value after L last random from a uniformly distributed sequence on the interval* [**L** *0,* **L** *1)* **c**;
l) **L real** *L last random* := **L** *.5*;

10.4. Synchronization Operations

a) **struct sema** = (**ref int** *F*);
b) **op** / = (**int** *a*) **sema** : (**sema** *s*; *F* **of** *s* := **int** := *a*; *s*);
c) **op** ↓ = (**sema** *edsger*) : (**ref int** *dijkstra* = *F* **of** *edsger*;
do (**if** *dijkstra* ≥ *1* **then** *dijkstra* −:= *1*; *l* **else c** *if the closed-statement replacing this comment is contained in a unitary-phrase which is a constituent unitary-phrase of the smallest collateral-phrase, if any, beginning with a parallel-symbol and containing this closed-statement, then the elaboration of that unitary-phrase is halted* {6.0.2.c}; *otherwise, the further elaboration is undefined* **c fi**); *l:* **skip**);
d) **op** ↑ = (**sema** *edsger*) : (**ref int** *dijkstra* = *F* **of** *edsger*;
dijkstra +:= *1*; **c** *the elaboration is resumed of all phrases whose elaboration is halted because the name possessed by dijkstra referred to a value smaller than one* **c**);

{See 2.2.5; for the use of ↓ and ↑, see E. W. Dijkstra, Cooperating Sequential Processes, contained in Programming Languages, Genuys, F. (ed.), London, etc., Academic Press, 1968; see also 11.13.}

10.5. Transput Declarations

{*"So it does!" said Pooh. "It goes in!"*
"So it does!" said Piglet. "And it comes out!"
"Doesn't it?" said Eeyore. "It goes in and out like anything."
Winnie-the-Pooh, *A. A. Milne.*}

10.5.0. Transput Modes and Straightening

10.5.0.1. Transput Modes

a) **union** % **simplout** = *(*⟨**L int**⟩, ⟨**L real**⟩, ⟨**L compl**⟩, **bool**, **char**, **string***)* ;

b) **union** % **outtype** = *(*⟨**D L int**⟩, ⟨**D L real**⟩, ⟨**D bool**⟩, ⟨**D char**⟩, ⟨**D outstruct**⟩*)* ;

c) **mode** % **outstruct** = **c** *an actual-declarer specifying a mode united from* {4.4.3.a} *all modes, except those specified by* **tamrof** *and by* **sema**, *which are structured from* {2.2.4.1.d} *only modes from which the mode specified by* **outtype** *is united* **c** ;

d) **struct** % **tamrof** = *(***string** *F1)* ; {See the remarks under 5.5.8.}

e) **union** % **intype** = *(*⟨**ref D L int**⟩, ⟨**ref D L real**⟩, ⟨**ref D bool**⟩, ⟨**ref D char**⟩, ⟨**ref D outstruct**⟩*)* ;

10.5.0.2. Straightening

a) **op** % **straightout** = *(***outtype** *x)* [ ] **simplout** :
 c *the result of "straightening" 'x'* **c**;

b) **op** % **straightin** = *(***intype** *x)* [ ] **intype** :
 c *the result of straightening 'x'* **c**;

c) The result of straightening a given value *V* is a multiple value obtained in the following steps:

Step 1: If *V* is (refers to) a value from whose mode that specified by **simplout** is united, then the result is a new instance of a multiple value composed of a descriptor consisting of an offset *1* and one quintuple *(1, 1, 1, 1, 1)* and *V* as its only element, and Step 4 is taken;

Step 2: If *V* is (refers to) a multiple value, then letting *n* stand for the number of elements of that value, and y_i for the result of straightening its *i*-th element, Step 3 is taken; otherwise, letting *n* stand for the number of fields of (of the value referred to by) *V*, and y_i for the result of straightening its *i*-th field, Step 3 is taken;

Step 3: If *V* is not (is) a name, then, letting m_i stand for the number of elements of y_i, the result is a new instance of a multiple value composed of a descriptor consisting of an offset *1* and one quintuple $(1, m_1 + \ldots + m_n, 1, 1, 1)$ and elements, the *l*-th of which, where $l = m_1 + \ldots + m_k - 1 + j$, is the (is the name referring to the) *j*-the element of y_k for $k = 1, \ldots, n$ and $j = 1, \ldots, m_k$;

Step 4: If *V* is not (is) a name, then the mode of the result is 'row of' followed by the mode specified by **simplout** (**intype**).

10.5.1. Channels and Files

{aa) "Channels", "backfiles" and files model the transput devices of the physical machine used in the implementation.

bb) A channel corresponds to a device, e.g., a card reader or punch, a magnetic drum or disc, to part of a device, e.g., a piece of core store, the keyboard of a teleprinter, or to a number of devices, e.g., a bank of tape units or even a set-up in nuclear physics the results of which are collected by the computer. A channel has certain properties (10.5.1.1.d to 10.5.1.1.o, table 1). A "random access" ("sequential access") channel is one for which the value of *set possible* (10.5.1.1.e) is **true** (**false**). The transput devices of some physical machine may be seen in more than one way as channels with properties. The choice made in an implementation is a matter for individual taste. Some possible choices are given in table 1.

cc) All information on a given channel is to be found in a number of backfiles. A backfile (10.5.1.1.b) comprises a three-dimensional array of integers (bytes of information), the *book* of the backfile; the lower bounds of the *book* are all **1**, the upper bounds are nonnegative integers, *maxpage, maxline* and *maxchar* of the backfile; furthermore, the backfile comprises the position of the "end of file", i.e., the page number, line number and char number up to which the backfile is filled with information, the current position and the "identification string" of the backfile.

dd) After the elaboration of the declaration of *chainbfile* (10.5.1.1.c), all backfiles form the chains of backfiles referenced by *chainbfile*, each backfile chained to the next one by its field *next*.

Examples:

a) In a certain implementation, channel **6** is a line printer. It has no input information, *chainbfile* [*6*] is initialized to refer to a backfile the *book* of which is an integer array with upper bounds **2000**, **60** and **144** (**2000** pages of continuous stationery), with both the current position and the end of file at **(1, 1, 1)** and *next* equal to **nil**. All elements of the *book* are left undefined.

b) Channel **4** is a drum, divided into **32** segments each being one page of **256** lines of **256** bytes. It has **32** backfiles of input information (the previous contents of the drum), so *chainbfile* [*4*] is initialized to refer to the first backfile of a chain of **32** backfiles, the last one having *next* equal to **nil**. Each of those backfiles has an end of file at position **(2, 1, 1)**.

c) Channel **20** is a tape unit. It can accommodate one tape at a time; one input tape is mounted and another tape laid in readiness. Here, *chainbfile* [*20*] is initialized to refer to a chain of two backfiles.

Since it is part of the standard declarations, all input is part of the program, though not of the particular-program.

ee) A file (10.5.1.2.a) is a structured value which comprises a reference to a backfile, and the information necessary for the transput routines to work with that backfile. A backfile is associated with a file by means of *open* (10.5.1.2.b), *create* (10.5.1.2.c) or *establish* (10.5.1.2.d). A given channel can accommodate a

Table 1. *Properties of some possible channels*

properties	card reader	card punch	magnetic tape unit			line printer
reset possible	*false*	*false*	*true*	*true*	*true*	*false*
set possible	*false*	*false*	*false*	*false*	*false*	*false*
get possible	*true*	*false*	*true*	*true*	*false*	*false*
put possible	*false*	*true*	*false*	*true*	*true*	*true*
bin possible	*false*	*true*	*false*	*true*	*false*	*false*
idf possible	*false*	*false*	*true*	*true*	*true*	*false*
reidf possible	*false*	*false*	*false*	*true*	*false*	*false*
max page	1	1	very large	very large	very large	very large
max line	large	very large	16	large	60	60
max char	72	80	84	large	144	144
stand conv	a 48- or 64-character code		64-char code	some code	line-pr code	line-pr code
max nmb files	1	1	1	1	1	1

properties	magnetic disc	magnetic drum		paper tape reader		tape punch
reset possible	*true*	*true*	*true*	*false*	*false*	*false*
set possible	*true*	*false*	*true*	*false*	*false*	*false*
get possible	*true*	*true*	*true*	*true*	*true*	*false*
put possible	*true*	*true*	*true*	*false*	*false*	*true*
bin possible	*true*	*true*	*true*	*false*	*true*	*false*
idf possible	*true*	*true*	*true*	*false*	*false*	*false*
reidf possible	*true*	*true*	*true*	*false*	*false*	*false*
max page	200	1	1	1	1	1
max line	16	1	256	very large	very large	very large
max char	128	524288	256	80	150	4
stand conv	some code	some code	some code	5-hole code	7-hole code	lathe code
max nmb files	10	4	32	1	1	1

certain number (10.5.1.1.o) of backfiles at any stage of the elaboration. The association is ended by means of *close* (10.5.1.2.s), *lock* (10.5.1.2.t) or *scratch* (10.5.1.2.u).

ff) When a file is "opened" on a channel for which the value of *idf possible* is **false**, then the first backfile is taken from the chain of backfiles for that channel, and is made the *bfile* of the file, obliterating the previous backfile, if any, of the file. When a file is opened on a channel for which the value of *idf possible* is **true**, then, if the given identification string is empty, then the first backfile, and, otherwise, the first backfile which has that identification string, is taken from the chain of backfiles for that channel; this backfile is made the *bfile* of the file.

gg) When a file is "established" on a channel, then a backfile is generated (8.5) with a *book* of the given size, the given identification string and both the current position and the end of file at *(1, 1, 1)*; when a file is "created" on a channel, then a file is established with a backfile the *book* of which has the maximum size for the channel and the *idf* of which is an empty string.

hh) When a file is "scratched", then its associated backfile is obliterated.

ii) When a file is "closed" ("locked"), then its associated backfile is attached to the chain referenced by *chainbfile* (*lockedbfile*) of the channel. Another file may (No file can) now be opened with this backfile by a call of *open*.

jj) The identification string of the backfile of a file opened on a channel for which the value of *reidf possible* is **true** may be changed by a call of *reidf*.

kk) A file comprises some fields of the mode 'procedure boolean', 'procedure with reference to character parameter boolean' or 'procedure with integral parameter boolean', routines which are called when in transput certain error situations arise. After opening or creating a file, the routines provided yield the value **false** when called, but the programmer may assign other routines to those fields. If the elaboration of such a routine is terminated, then the transput routine which called it can take no further action; otherwise, if it yields the value **true**, then it is assumed that the error situation has been remedied in some way, and, if possible, transput goes on, but if it yields the value **false**, then *undefined* is called, i.e., some sensible system action is taken (rr). These routines are:

a) *logical file end*, which is called when during input from a file on a sequential channel the end of file of its backfile is passed. If the routine yields the value **true**, then transput goes on, and if it yields **false**, then some sensible action is taken.

Example:

The programmer wishes to count the number of integers on his input tape. The file *intape* was opened in a surrounding range. If he writes

```
begin int n := 0; logical file end of intape := go to f;
      do (get (intape, loc int); n +:= 1); f: print (n)
end ,
```

then the assignment to the field of *intape* violates the scope restrictions (; the scope of the routine *((*: **go to** *f))* is smaller than the scope of *intape*), so he has to write

```
begin int n := 0; file auxin := intape;
      logical file end of auxin := go to f;
      do (get (auxin, loc int); n +:= 1); f: print (n)
end .
```

b) *physical file end*, which is called when the *max page*, the *max line* or the *max char* of the backfile of a file is exceeded. If the routine yields the value *true*, then transput goes on, and if it yields *false*, then some sensible action is taken.

Example:

The programmer wishes automatically to give a new line at the end of a line and a new page at the end of a page on his file *f*:

```
proc bool new line page = bool : ((line ended (f) | new line (f));
  (page ended (f) | new page (f)); true);
```

c) *char error*, which is called when, during formatted input, a character is read which does not agree with the frame specifying it (5.5.1.m) or when, during input, at the current position an uninterpretable character is present (10.5.1.ll), with a reference to a character, suggested as a replacement. The routine provided by the programmer may give some other character than the suggested one. If the routine yields *true*, then that suggested character as possibly modified by the routine is used, and, if it yields *false*, then some sensible action is taken.

Example:

The programmer wishes to print a list of all such disagreements. He assigns to the field *char error* of his file *f*

```
((ref char sugg) bool : (char k; backspace (f);
  int p = page number (f), l = line number (f), c = char number (f);
  get (f, k); print ((new line, "at", p, l, c, "present.""", k, """",
  .suggested."""", sugg, """"."")); true));
```

d) *value error*, which is called when, during formatted transput, an attempt is made to transput a value under control of a picture with which it is not compatible, or when the number of frames is not sufficient. If the routine yields *true*, then the current value and picture are skipped, i.e., transput goes on at 5.5.1.dd.Step 5; if the routine yields *false*, then first, on output, the value is output by *put*, and next some sensible action is taken.

e) *format end*, which is called when, during formatted transput, the format is exhausted while still some value remains to be transput. If the routine yields *true*, then transput goes on (so the routine must have provided a new format for the file), and, if the routine yields *false*, then the current format is repeated, i.e., the first picture again is made to be the current picture of the file.

f) *other error*, which is called with some actual-integral-parameter, when during transput some other error situation arises. No call of this routine occurs explicitly in the standard-prelude, and neither the meaning of its actual-parameter nor that of the value yielded, is defined in this Report.

This routine may, in some implementation, be called when an incorrigible hardware error occurs which makes transput involving this file impossible. (The programmer may provide a routine which then closes the file and opens it on some other channel.)

ll) The *conv* of a file is used by the transput routines in the conversion of characters to and from integers in the *book* of the *bfile* of the file. After opening, creating or establishing a file, *stand conv* of the channel is used, but some other "conversion key" may be provided by the programmer by a call of *make conv* (10.5.1.2.z).

On output, the given character is converted to that integer, if any, in the conversion key, whose ordinal number is the integral equivalent of that character; what action is taken when an attempt is made to convert a character with an integral equivalent exceeding the upper bound of the conversion key, is left undefined; on input, the given integer is converted to that character, if any, whose integral equivalent is the lowest ordinal number for which the element of the conversion key is equal to that given integer; if no such character exists, then *char error* is called with a space (parity error, nonexistent code).

mm) The *term* of a file is used in reading strings of a variable number of characters, where any of the characters of *term* serves as a terminator (see 5.5.1.jj and 10.5.2.2.dd). This terminator string may be provided by the programmer. Furthermore, when reading outside the file, *physical file end* of the file is called, and if that does not cause the position to be within the file, that also serves as a terminator.

nn) On a channel for which the value of *reset possible* is **true**, a file may be "reset", causing its position to be *(1, 1, 1)*. On a sequential access file, the end of file remains at the position up to which the backfile contains information, but when after resetting any output is done, then the end of file is first set at the current position.

oo) On a random access channel, a file may be "set", causing its position to be the given position.

pp) On files opened on a sequential access channel, binary and nonbinary transput may not be alternated, i.e., after opening, creating or resetting such a file, either is possible, but, once one has taken place on the file, the other may not until the file has been reset again.

qq) On files opened on a sequential access channel for which *put possible* and *get possible* both possess the value **true**, nonbinary input and output may be alternated, but it is not allowed to read past the end of file.

rr) When in transput something happens which is left undefined, for instance by an explicit call of *undefined* (10.5.1.2.y), this does not imply that the elaboration is catastrophically and immediately terminated, but only that some sensible action is taken which is not or cannot be described by this Report alone, and is generally implementation dependent. For instance, in some implementation it may be possible to set a tape unit to any position within the logical file, even if the value of *set possible* is **false** (oo).

Example:

begin file *f1, f2;* [*1* : *10000*] **int** *x;* **int** *n; open (f1, , channel 2);*
f2 := *f1;* # *now f1 and f2 can be used interchangeably* #
make conv (f1, flexocode); make conv (f2, telexcode);
now f1 and f2 use different codes; flexocode and telexcode are defined in the library-declaration for this implementation # *reset (f1);*
consequently, f2 is reset too
for *i* **while** ¬ *logical file ended (f1)* **do** *(n* := *i; get (f1, x* [*i*]*));*
too bad if there are more than 10000 integers in the input # *reset (f1);*
for *i* **to** *n* **do** *put (f2, x* [*i*]*); reset (f2); close (f2)* # *f1 is now closed too* #
end }

10.5.1.1. Channels

a) **int** *nmb channels* = **c** *an integral-clause indicating the number of transput channels in the implementation* **c**;

b) **struct** % **bfile** = *(*[*1* : *0* **flex**, *1* : *0* **flex**, *1* : *0* **flex**] **int** *book,*
int *lpage, lline, lchar, page, line, char, max page, max line, max char,*
string *idf,* **ref bfile** *next)*;

c) [*1* : *nmb channels*] **ref bfile** % *chainbfile* := **c** *some appropriate initialization* {see 10.5.1.dd} **c**;

d) [*1* : *nmb channels*] **bool** *reset possible* = **c** *a row-of-boolean-clause, indicating which of the physical devices corresponding to the channels allow resetting* {e.g., rewinding of a magnetic tape} **c**;

e) [*1* : *nmb channels*] **bool** *set possible* = **c** *a row-of-boolean-clause, indicating which devices can be accessed at random* **c**;

f) [*1* : *nmb channels*] **bool** *get possible* = **c** *a row-of-boolean-clause, indicating which devices can be used for input* **c**;

g) [*1* : *nmb channels*] **bool** *put possible* = **c** *a row-of-boolean-clause, indicating which devices can be used for output* **c**;

h) [*1* : *nmb channels*] **bool** *bin possible* = **c** *a row-of-boolean-clause, indicating which devices can be used for binary transput* **c**;

i) [*1* : *nmb channels*] **bool** *idf possible* = **c** *a row-of-boolean-clause, indicating on which devices backfiles have an identification* **c**;

j) [*1* : *nmb channels*] **bool** *reidf possible* = **c** *a row-of-boolean-clause, indicating on which devices backfiles allow reidentification* **c**;

k) [*1* : *nmb channels*] **int** *max page* = **c** *a row-of-integral-clause, giving the maximum number of pages per file for the channel* **c**;

l) [*1* : *nmb channels*] **int** *max line* = **c** *a row-of-integral-clause, giving the maximum number of lines per page* **c**;

m) [*1* : *nmb channels*] **int** *max char* = **c** *a row-of-integral-clause, giving the maximum number of characters per line* **c**;

n) [*1* : *nmb channels*] **struct** *(***proc** [ ] **int** *F) stand conv* = **c** *a clause giving the standard conversion keys for the channels* {; other conversion keys may be provided by the library-prelude} **c**;

o) [*1* : *nmb channels*] **int** *max nmb files* = **c** *a row-of-integral-clause, giving the maximum number of files the channels can accommodate* **c**;

```
p) [1 : nmb channels] int % nmb opened files;
     for i to nmb channels do nmb opened files [i] := 0;
q) [1 : nmb channels] ref bfile % lockedbfile;
     for i to nmb channels do lockedbfile [i] := nil;
r) proc file available = (int channel) bool :
     nmb opened files [channel] < max nmb files [channel];
```

10.5.1.2. Files

```
a) struct file = (ref bfile % bfile, int % chan, ref int % forp,
     ref bool % state def, % state get, % state bin, % opened,
     ref string % format, string term, [0 : 0 flex] int % conv,
     proc bool logical file end, physical file end, format end, value error,
     proc (ref char) bool char error, proc (int) bool other error);
b) proc open = (ref file file, string idf, int ch) :
     if file available (ch)
     then ref ref bfile bf := chainbfile [ch];
       while (ref bfile : bf) :≠: nil do
         (idf of bf = idf ∨ idf = "" ∨ ¬ idf possible [ch] |
           l | bf := next of bf); undefined.
     l: file := (bf, ch, int := 0, bool := false, bool, bool, bool := true,
                  nil, "", F of standconv [ch], false, false, false, false,
                  ((ref char a) bool : false), skip);
       (ref ref bfile : bf) := next of bf; nmb opened files [ch] +:= 1
     else undefined fi;
c) proc create = (ref file file, int ch) :
     establish (file,, max page [ch], max line [ch], max char [ch], ch);
d) proc establish = (ref file file, string idf, int mp, ml, mc, ch) :
     if file available (ch) ∧ mp ≤ max page [ch] ∧ ml ≤ max line [ch] ∧
       mc ≤ max char [ch]
     then bfile bf = ([1 : mp, 1 : ml, 1 : mc] int, 1, 1, 1, 1, 1, 1, mp, ml, mc,
                        idf, nil);
       file := (bfile := bf, ch, int := 0, bool := false, bool,
                  bool, bool := true, nil, "", F of standconv [ch], false,
                  false, false, false, ((ref char a) bool : false), skip);
       nmb opened files [ch] +:= 1
     else undefined fi;
e) proc set = (file file, int p, l, c) :
     if set possible [chan of file] ∧ opened of file
     then page of bfile of file := p; line of bfile of file := l;
          char of bfile of file := c; check plc (file)
     else undefined fi;
f) proc reset = (file file) :
     if reset possible [chan of file] ∧ opened of file
     then page of bfile of file := 1; line of bfile of file := 1;
          char of bfile of file := 1; state def of file := false
     else undefined fi;
```

```
g) proc % check plc = (file file) : if opened of file
      then (¬ (logical file ended (file) | logical file end of file |:
               line ended (file) ∨ page ended (file) ∨ file ended (file) |
               physical file end of file | true) | undefined)
      else undefined fi;
h) proc line ended = (file file) bool : (opened of file |
      int c = char of bfile of file; c ≤ 0 ∨ c > max char of bfile of file);
i) proc page ended = (file file) bool : (opened of file |
      int l = line of bfile of file; l ≤ 0 ∨ l > max line of bfile of file);
j) proc file ended = (file file) bool : (opened of file |
      int p = page of bfile of file; p ≤ 0 ∨ p > max page of bfile of file);
k) proc logical file ended = (file file) bool : (opened of file |:
      ¬ set possible [chan of file] ∧ state def of file ∧ state get of file |
      bfile b = bfile of file; int p = page of b, lp = lpage of b,
        l = line of b, ll = lline of b, c = char of b, lc = lchar of b;
      (p < lp | false |: p > lp | true |: l < ll | false |: l > ll | true | c ≥ lc) | false);
l) proc % get string = (file file, ref [1 : either] char s) :
     if get possible [chan of file] ∧ opened of file
     then ref int p = page of bfile of file, l = line of bfile of file,
                  c = char of bfile of file;
       if ¬ set possible [chan of file] thef state def of file
       then (state bin of file | undefined) fi;
       state def of file := state get of file := true;
       state bin of file := false;
       for i to ⌈ s do (check plc (file);
         for j from 0 to ⌈ conv of file do
           ((conv of file) [j] = (book of bfile of file) [p, l, c] |
            s [i] := repr j; e);
         char k := "."; s [i] := ((char error of file) (k) | k | undefined; ".");
         e: c +:= 1)
     else undefined fi;
m) proc % put string = (file file, string s) :
     if put possible [chan of file] ∧ opened of file
     then ref int p = page of bfile of file, l = line of bfile of file,
                  c = char of bfile of file;
       if ¬ set possible [chan of file] thef state def of file
       then (state bin of file | undefined) fi;
       state get of file := state bin of file := false;
       state def of file := true;
       for i to ⌈ s do (check plc (file);
         (book of bfile of file [p, l, c] := (conv of file) [abs s [i]]; c +:= 1;
         (p = lpage of bfile of file ∧ l = lline of bfile of file |
          (c > lchar of bfile of file | lchar of bfile of file := c) |
          lpage of bfile of file := p; lline of bfile of file := l;
          lchar of bfile of file := c))
     else undefined fi;
```

n) **proc** *char in string* = (**char** *c*, **ref int** *i*, **string** *s*) **bool** :
  (**for** *k* **to** ⌈ *s* **do** (*c* = *s* [*k*] | *i* := *k*; *l*); **false**. *l*: **true**);

o) **proc** *space* = (**file** *file*) : (*char* **of** *bfile* **of** *file* +:= *1*; *check plc* (*file*));

p) **proc** *backspace* = (**file** *file*) : (*char* **of** *bfile* **of** *file* −:= *1*; *check plc* (*file*));

q) **proc** *new line* = (**file** *file*) : (*line* **of** *bfile* **of** *file* +:= *1*;
  char **of** *bfile* **of** *file* := *1*; *check plc* (*file*));

r) **proc** *new page* = (**file** *file*) : (*page* **of** *bfile* **of** *file* +:= *1*;
  line **of** *bfile* **of** *file* := *char* **of** *bfile* **of** *file* := *1*; *check plc* (*file*));

s) **proc** *close* = (**file** *file*) : (*opened* **of** *file* | **int** *ch* = *chan* **of** *file*;
  next **of** *bfile* **of** *file* := *chainbfile* [*ch*]; *chainbfile* [*ch*] := *bfile* **of** *file*;
  opened **of** *file* := **false**; *nmb opened files* [*ch*] −:= *1*);

t) **proc** *lock* = (**file** *file*) : (*opened* **of** *file* | **int** *ch* = *chan* **of** *file*;
  ref bfile *bf* = *bfile* **of** *file*; *page* **of** *bf* := *line* **of** *bf* := *char* **of** *bf* := *1*;
  next **of** *bf* := *lockedbfile* [*ch*]; *lockedbfile* [*ch*] := *bf*;
  opened **of** *file* := **false**; *nmb opened files* [*ch*] −:= *1*);

u) **proc** *scratch* = (**file** *file*) : (*opened* **of** *file* |
  opened **of** *file* := **false**; *nmb opened files* [*chan* **of** *file*] −:= *1*);

v) **proc** *char number* = (**file** *f*) **int** : (*opened* **of** *f* | *char* **of** *bfile* **of** *f*);

w) **proc** *line number* = (**file** *f*) **int** : (*opened* **of** *f* | *line* **of** *bfile* **of** *f*);

x) **proc** *page number* = (**file** *f*) **int** : (*opened* **of** *f* | *page* **of** *bfile* **of** *f*);

y) **proc** % *undefined* = **c** *some sensible system action* {10.5.1.rr} **c**;

z) **proc** *make conv* = (**ref file** *f*, **struct** (**proc** [ ] **int** *F*) *c*) :
  conv **of** *f* := *F* **of** *c*;

aa) **proc** *reidf* = (**file** *f*, **string** *idf*) :
  (*reidf possible* [*chan* **of** *f*] | *idf* **of** *bfile* **of** *f* := *idf*);

10.5.1.3. Standard Channels and Files

a) **int** *stand in channel* = **c** *an integral-clause such that the value of get possible* [*stand in channel*] *is true* **c**;

b) **int** *stand out channel* = **c** *an integral-clause such that the value of put possible* [*stand out channel*] *is true* **c**;

c) **int** *stand back channel* = **c** *an integral-clause such that the values of reset possible* [*stand back channel*], *set possible* [*stand back channel*], *get possible* [*stand back channel*], *put possible* [*stand back channel*], *and bin possible* [*stand back channel*] *are true* **c**;

d) **file** % *f*; *open* (*f*,, *stand in channel*); **file** *stand in* = *f*;

e) *open* (*f*,, *stand out channel*); **file** *stand out* = *f*;

f) *open* (*f*,, *stand back channel*); **file** *stand back* = *f*;

{Certain "standard files" (d, e, f) need not (and cannot) be opened by the programmer, but are opened for him in the standard-prelude; *print* (10.5.2.1.a) can be used for output on *stand out*, *read* (10.5.2.2.a) for input from *stand in*, and *write bin* (10.5.4.1.a) and *read bin* (10.5.4.2.a) for transput involving *stand back*. The programmer need not close these standard files, since they are locked in the standard-postlude.}

10.5.2. Formatless Transput

10.5.2.1. Formatless Output

{For formatless output, *print* and *put* can be used. The elements of the given value of the mode specified by [ ] **union** *(***outtype**, **proc** *(***file***))* are treated one after the other; if an element is of the mode specified by **proc** *(***file***)* (i.e., a "layout procedure"), then it is called with the file as its parameter; otherwise, it is straightened (10.5.0.2), and the resulting values are output on the given file one after the other, as follows:

aa) If the mode of the value is specified by **L int**, then first, if there is not enough room on the line for *L int width* + *2* characters, then this room is made by giving a new line and, if the page is full, giving a new page; next, when not at the beginning of a line, a space is given and the value is edited as if under control of the picture *n(L int width* − *1)z+d*.

bb) If the mode of the value is specified by **L real**, then first, if there is not enough room on the line for *L real width* + *L expwidth* + *5* characters, then this room is made; next, when not at the beginning of a line, a space is given, and the value is edited as if under control of the picture *+d.n(L real width* − *1) den(L expwidth* − *1)z+d*.

cc) If the mode of the value is specified by **L compl**, then first, if there is not enough room on the line for *2* × *(L real width* + *L exp width)* + *11* characters, then this room is made; next, when not at the beginning of a line, a space is given, and the value is edited as if under control of the picture *+d.n(L real width* − *1)den(L expwidth* − *1)z+d"."i+d.n(L real width* − *1)den(L expwidth* − *1)z+d*.

Table 2. Display of the values of *L int string*, *L dec string* and *L real string* (10.5.2.1.c,e,d)

	w − 1
L int string:	+DDDDDDDDDDD
	w
	w − d − 2 d
L dec string:	+DDDDDDD.DDDDD
	w
	w−d−e−4 d e
L real string:	+DDDDDDD.DDDDD$_{10}$+DDD
	w

Table 3. Significance of the elements of *frame* (10.5.3.i)

frame	
[*1*]	type (*1* = integer, *2* = real fixed, *3* = real floating, *4* = complex fixed, *5* = complex floating, *6* = string, *7* = integer choice, *8* = boolean, *9* = character);
[*2*]	radix (*2*, *4*, *8*, *10* or *16*);
[*3*]	sign (*0* = no sign frame, *1* = sign frame +, *2* = sign frame −);
[*4*]	number of digits before point; if type = *1*, then w − *1*, else if type = *2* or *4*, then w − d − *2*, else if type = *3* or *5*, then w − d − e − *4*, or, if type = *6*, then the number of characters in the string if this is constant, or *0*, if this number is variable;
[*5*]	number of digits after point; if type = *2*, *3*, *4* or *5* then d;
[*6*]	sign of exponent; if type = *3* or *5*, then as [*3*];
[*7*]	number of digits in exponents; if type = *3* or *5*, then e;
[*8*], ..., [*14*] as [*1*], ..., [*7*] when *frame* [*1*] = *4* or *5*.	

dd) If the mode of the value is specified by [ ] **char**, then its elements are written one after the other.

ee) If the mode of the value is specified by **char**, then first, if the line is full, then room is made, and next the character is written.

ff) If the mode of the value is specified by **bool**, then, if the value is *true* (*false*), then the character possessed by the flip- (flop-)symbol is output as ee.}

a) **proc** *print* = ([ ] **union** (**outtype**, **proc** (**file**)) *x*) : *put* (*stand out*, *x*);

b) **proc** *put* = (**file** *file*, [*1* :] **union** (**outtype**, **proc** (**file**)) *x*) :

```
    begin outtype ot; proc (file) pf;
         for i to ⌈ x do (ot ::= x [i]; pf ::= x [i] | pf (file) |
           [1 : ] simplout y = straightout ot;
           for j to ⌈ y do (string s, bool b, char k;
             ((<(L int i; (i ::= y [j] |
                  s := L int string (i, L int width + 1, 10);
                  sign supp zero (s, 1, L int width)))>);
             ((<(L real x; (x ::= y [j] | s := L real conv (x)))>);
             ((<(L compl z; (z ::= y [j] |
                  s := L real conv (re z) + ".⊥" + L real conv (im z)))>);
             (b ::= y [j] | put (file, (b | "1" | "0")); end);
             (k ::= y [j] | nextplc (file); put string (file, k); end);
             (s ::= y [j] | put string (file, s); end);
             ref int c = char of bfile of file; int c1 = c, n = ⌈ s;
             c +:= (c1 = 1 | n | n + 1);
             (line ended (file) | nextplc (file) | c := c1);
             put string (file, (c = 1 | s | "." + s));
         end: skip))
    end;
```

```
c) proc L int string = (L int x, int w, r) string : (r>1∧r>17 |
      string c := ; L int n := abs x; L int lr = K r;
      for i to w − 1 do (dig char (S (n ÷: lr)) +=: c; n ÷:= lr);
      (n = L 0 | (x ≥ L 0 | "+" | "−") + c | "") | "");
d) proc L real string = (L real x, int w, d, e) string :
      (d≥0 ∧ e>0 ∧ d+e+4≤w |
       L real g = L 10 ↑ (w − d − e − 4); L real h = g × L .1;
       L real y := abs x; int p := 0; while y≥g do (y ×:= L .1; p +:= 1);
       (y> L 0 | while y<h ∧ p − 1 > − (10 ↑ e) do (y ×:= L 10; p −:= 1));
       (y + L .5 × L .1 ↑ d ≥ g | y := h; p +:= 1);
       L dec string ((x≥0 | y | −y), w − e − 2, d) + "10" +
         int string (p, e + 1, 10));
e) proc L dec string = (L real x, int w, d) string :
      (abs x < L 10 ↑ (w − d − 2) ∧ d≥0 ∧ d+2≤w | string s := ;
       L real y := (abs x + L .5 × L .1 ↑ d) × L .1 ↑ (w − d − 2);
       to w − 2 do s +:= dig char ((int c = S ⌊ (y ×:= L 10); y −:= K c; c));
       (x≥0 | "+" | "−") + s [1 : w − d − 2] + "." + s [w − d − 1 : ]);
f) proc % dig char = (int x) char : ("0123456789abcdef" [x + 1]);
```

{In connection with 10.5.2.1.c,d,e, see Table 2.}

```
g) proc % sign supp zero = (ref string c, int l, u) :
      for i from l + 1 to u while c [i] = "0" do
        (c [i] := c [i − 1]; c [i − 1] := ".");
h) int L int width = (int c := 1;
      while L 10 ↑ (c − 1) < L .1 × L max int do c +:= 1; c);
i) int L real width = 1 − S ⌊ (L ln (L small real) / L ln (L 10));
j) int L exp width =
      1 + S ⌊ (L ln (L ln (L max real) / L ln (L 10)) / L ln (L 10));
k) proc % L real conv = (L real x) string :
      (string s := L real string (x, L real width + L exp width + 4,
        L real width − 1, L exp width); sign supp zero (s, L real width + 4,
        L real width + L exp width + 3); s);
l) proc % next plc = (file file) : (opened of file |
      (line ended (file) | new line (file)); (page ended (file) | new page (file)));
```

10.5.2.2. Formatless Input

{For formatless input, *read* and *get* can be used. The elements of the given value of the mode specified by [ ] **union (intype, proc (file))** are treated one after the other; if an element is a layout procedure, then it is called with the file as its parameter; otherwise, it is straightened (10.5.0.2), and to the resulting names values are assigned, input from the given file as follows:

aa) If the name refers to a value whose mode is specified by **L int**, then first, the file is searched for the first character that is not a space (giving new lines and pages as necessary); next, the largest string is read from the file that could be indited under control of some picture of the form *n(k2)dd* or *+n(k1)"." n(k2)dd*; this string is converted to an integer by *L string int*.

bb) If the name refers to a value whose mode is specified by **L real**, then first, the file is searched for the first character that is not a space; next, the largest string is read from the file that could be indited under control of a picture of the form *+n(k1)"."n(k2)d* or *n(k2)d* followed by *.n(k3)dd* or *ds.* possibly followed by *en(k4)"."+n(k5)"."n(k6)dd* or *en(k5)"."n(k6)dd*; this string is converted to a real number by *L string real*.

cc) If the name refers to a value whose mode is specified by **L compl**, then first, a real number is input as in bb, and assigned to the real part; next, the file is searched for the first character that is not a space; next, a plus i times is expected; finally, a real number is input and assigned to the imaginary part.

dd) If the name refers to a value whose mode is specified by [ ] **char**, then, if both upper and lower state of the value are *1*, then as many characters are read as the value has elements; if not both states are *1*, then characters are read from the line under control of the terminator string referenced by the file (5.5.1.jj, 10.5.1.mm); the string with those characters as its elements is then the resulting value.

ee) If the name refers to a value whose mode is specified by **char**, then first, if the line is full, a new line is given, and, if the page is full, a new page is given; next, the character is read from the file.

ff) If the name refers to a value whose mode is specified by **bool**, then first, the file is searched for the first character that is not a space; then a character is read; if this character is that possessed by the flip-(flop-)symbol, then the resulting value is *true* (*false*); if the character is neither of those, then *char error* of the file is called with *false*.}

```
a) proc read = ([ ] union (intype, proc (file)) x) : get (stand in, x);
b) proc get = (file file, [1 : ] union (intype, proc (file)) x) :
     begin intype it; proc (file) pf; char k; priority ! = 8;
           for i to ⌈ x do (it ::= x [i]; pf ::= x [i] | pf (file) |
             [1 : ] intype y = straightin it;
             op ? = (string s) bool : (outside (file) | false | :
               get string (file, k); char in string (k, loc int, s) |
               true | backspace (file); false);
             op ? = (char c) bool : ? (string : c);
             op ! = (string s, char c) char : (get string (file, k);
               char in string (k, loc int, s) | k | char sugg := c;
               ((char error of file) (sugg) | sugg | undefined; c));
             op ! = (char s, c) char : (string : s) ! c;
             proc skip spaces = : while (next plc (file); ? ".") do skip;
             proc string read dig = string : (string t := "0123456789" ! "0";
               while ? "0123456789" do t +:= k; t);
             proc string read num = string :
               (char t := (skip spaces; ? "+−" | k | "+");
                 while ? "." do skip; t + read dig);
             proc string read real = string : (string t := read num;
               (? "." | t +:= "." + read dig);
                 (? "e10" | t +:= "10" + read num); t);
```

```
          for j to ⌈ y do (ref bool bb; ref char cc; ref string ss;
            (⟨(ref L int ii; (ii ::= y [j] |
              (ref L int : ii) := L string int (read num, 10)))⟩);
            (⟨(ref L real xx; (xx ::= y [j] |
              (ref L real : xx) := L string real (read real)))⟩);
            (⟨(ref L compl zz; (zz ::= y [j] | get (file, re of zz);
              skip spaces; "⊥" ! "⊥"; get (file, im of zz)))⟩);
            (bb ::= y [j] | skip spaces;
              (ref bool : bb) := (? "1" | true | "0" ! "0" = "1"));
            (cc ::= y [j] | next plc (file); get string (file, cc));
            (ss ::= y [j] |: ⌊ ss ∧ ⌈ ss | get string (file, ss [@ 1]) |
              string t; while ((outside (file) |
                physical file end of file); outside (file) | false |:
                ? term of file | backspace (file); false | true) do t +:= k;
                (ref string : ss) :=
                  t [@ (⌊ ss | ⌊ ss |: ⌈ ss | 1 − ⌈ t + ⌈ ss | 1)])))
    end;
```

c)
```
proc L string int = (string x, int r) L int :
  (r>1 ∧ r<17 | L int n := L 0; L int lr = K r; int w = ⌈ x;
  for i from 2 to w do n := n × lr + K (int d = char dig (x [i]);
  (d<r | d)); (x [1] = "+" | n |: x [1] = "−" | −n));
```

d)
```
proc L string real = (string x) L real :
  (int e; ((char in string ("₁₀", e, x) | true | char in string ("e", e, x)) |
    L string dec (x [1 : e − 1]) × L 10.0 ↑ string int (x [e + 1 : ], 10) |
    L string dec (x)));
```

e)
```
proc L string dec = (string x) L real : (int w = ⌈ x; L real r := L 0;
  int p; (char in string (".", p, x) |
    [1 : w − 2] char s = x [2 : p − 1] + x [p + 1 : ];
    for i to w − 2 do r := L 10 × r + K (int d = char dig (s [i]);
      (d<10 | d)); (x [1] = "+" | r |: x [1] = "−" | −r) ×
       L .1 ↑ (w − p) | L string dec (x + ".")));
```

f)
```
proc % char dig = (char x) int : (int i;
  (char in string (x, i, "0123456789abcdef") | i − 1 | undefined; 0));
```

g)
```
proc % outside = (file file) bool :
  line ended (file) ∨ page ended (file) ∨ file ended (file);
```

10.5.3. Formatted Transput

{For the significance of formats see format-denotations (5.5).}

a)
```
proc format = (file file, tamrof tamrof) :
  (forp of file := 1; format of file :=
    collection list pack ("(" + F1 of tamrof + ")", loc int := 1));
```

b)
```
proc % collection list pack = (string s, ref int p) string :
  (string t := collection (s, p);
    while s [p] = "," do t +:= "," + collection (s, p); p +:= 1; t);
```

```
c) proc % collection = (string s, ref int p) string :
     (int n, q; string f := (p +:= 1; insertion (s, p));
       q := p; replicator (s, p, n);
       (s [p] = "(" | string t = collection list pack (s, p);
         to n do f +:= t | p := q; f +:= picture (s, p, loc [1 : 14] int));
         f + insertion (s, p)) ;
d) proc % insertion = (string s, ref int p) string :
     (int q = p; skip insertion (s, p); s [q : p - 1]) ;
e) proc % skip insertion = (string s, ref int p) :
     while (p > ⌈ s | false |: skip align (s, p) | true | skip lit (s, p)) do skip;
f) proc % skip align = (string s, ref int p) bool :
     (int q = p; replicator (s, p, loc int);
       (char in string (s [p], loc int, "xyplk") |
         p +:= 1; true | p := q; false)) ;
g) proc % replicator = (string s, ref int p, n) :
     (string t; while char in string (s [p], loc int, "0123456789") do
       (t +:= s [p]; p +:= 1); n := (t = "" | 1 | string int ("+" + t, 10))) ;
h) proc % skip lit = (string s, ref int p) bool :
     (int q = p; replicator (s, p, loc int);
       (s [p] = """" | while (s [p +:= 1] = """" | s [p +:= 1] = """" |
         true) do skip; true | p := q; false)) ;
i) proc % picture = (string format, ref int p, ref [ ] int frame) string :
     begin int n; int po = p; bool a;
           op ? = (string s) bool :
             (skip insertion (format, p); p > ⌈ format | false |
               int q = p; replicator (format, p, n); a := q = p;
               (format [p] = "s" | p +:= 1);
               (char in string (format [p], loc int, s) |
                 p +:= 1; true | p := q; false)) ;
           op ? = (char c) bool : ? (string : c) ;
           proc intreal pattern = (ref [1 : 7] int frame) bool :
             ((num mould (frame [2 : 4]) | frame [1] := 1; l);
              (? "." |: num mould (frame [3 : 5]) | frame [1] := 2; l);
              (? "e" |: num mould (frame [5 : 7]) | frame [1] := 3; l);
              false. l: true);
           proc num mould = (ref [1 : 3] int frame) bool :
             ((? "r" | frame [1] := n); (? "z" | frame [3] +:= n);
              (? "+" | frame [2] := 1 |: ? "-" | frame [2] := 2);
              while ? "dz" do frame [3] +:= n;
              format [p] = "," ∨ format [p] = "i" ∨ format [p] = ")") ;
           proc string mould = (ref [ ] int frame) bool : (? "t" | true |
             while ? "a" do frame [4] +:= n; format [p] = "," ∨
               format [p] = ")") ;
           for i to 14 do frame [i] := 0; frame [2] := 10;
           (intreal pattern (frame [1 : 7]) | (? "i" |
             frame [1] +:= 2; intreal pattern (frame [8 : 14])); end) ;
```

```
            (string mould (frame) | frame [1] := (frame [4] = 1 ∧ a | 9 | 6);
             end); (? "b" | frame [1] := 8 |: ? "c" | frame [1] := 7 |
               frame [1] := 0; end);
            (format [p] = "(" | while ? "(," do skip lit (format, p); p +:= 1);
     end: skip insertion (format, p); format [po : p - 1]
  end;
```

{In connection with 10.5.3.i see Table 3.}

10.5.3.1. Formatted Output

a)
```
  proc outf = (file file, tamrof tamrof, [ ] outtype x) :
     (format (file, tamrof); out (file, x));
```
b)
```
  proc out = (file file, [1 : ] outtype x) :
     begin string format = format of file; ref int p = forp of file;
           for k to ⌈ x do
           ([1 : ] simplout y = straightout x [k]; int q, j := 0;
            [1 : 14] int frame;
       rep: j +:= 1;
      step: while (do insertion (file, format, p); p > ⌈ format | false |
           format [p] = ",") do p +:= 1; (j > ⌈ y | end);
           (p > ⌈ format | (¬ format end of file | p := 1); step);
           q := p; picture (format, q, frame);
           (frame [1] | int, real, real, compl, compl, string, intch, bool, char);
       int: (<(L int i; (i ::= y [j] |
            edit L int (file, i, format, p, frame); rep))>); incomp;
      real: (<(L real x; (x ::= y [j] |
            edit L real (file, x, format, p, frame); rep))>);
           (<(L int i; (i ::= y [j] |
            edit L real (file, i, format, p, frame); rep))>); incomp;
     compl: (<(L compl z; (z ::= y [j] |
           edit L compl (file, z, format, p, frame); rep))>);
           (<(L real x; (x ::= y [j] |
            edit L compl (file, x, format, p, frame); rep))>);
           (<(L int i; (i ::= y [j] |
            edit L compl (file, i, format, p, frame); rep))>); incomp;
    string: ([1 flex : 0 flex] char s; (s ::= y [j] |: frame [4] = 0 |
            put (file, s) | edit string (file, s [@ 1], format, p, frame); rep));
      char: (char c; (c ::= y [j] |
            edit string (file, c, format, p, frame); rep)); incomp;
     intch: (int i; (i ::= y [j] | edit choice (file, i, format, p); rep)); incomp;
      bool: (bool b; (b ::= y [j] | edit bool (file, b, format, p); rep));
    incomp: (value error of file | rep | put (file, y [j]); undefined);
       end: skip)
       end;
```
c)
```
  proc % edit L int = (file f, L int i, string format, ref int p, [ ] int fr) :
     (string s = L int string (i, fr [4] + 1, fr [2]);
       (s = "" | (¬ value error of f | put (f, i); undefined) |
         edit string (f, s, format, p, fr)));
```

```
d) proc % edit L real = (file f, L real x, string format, ref int p, [ ] int fr):
     (string s = stringed L real (x, fr); int t := −1;
       (¬ char in string ("₁₀", t, s) | char in string ("e", t, s));
         (t = ⌈ s | (¬ value error of f | put (f,i); undefined) |
           edit string (f, s, format, p, fr)));

e) proc % stringed L real = (L real x, [ ] int fr) string:
     (fr [1] = 2 | L dec string (x, fr [4] + fr [5] + 2, fr [5]) |
       L real string (x, fr [4] + fr [5] + fr [7] + 4, fr [5], fr [7]));

f) proc % edit L compl = (file f, L compl z, string format, ref int p, [ ] int fr):
     edit string (f, ([1 : 14] int g := fr; g [1] −:= 2; stringed L real
       (re z, g [1 : 7]) + "⊥" + stringed L real (im z, g [8 : 14])),
         format, p, fr);

g) proc % edit string = (file f, string x, format, ref int p, [ ] int fr):
     begin int p1 := 1, n; bool supp; string s := x;
           op ? = (char s) bool: (do insertion (f, format, p); p > ⌈ format |
              false | int q = p; replicator (format, p, n);
              (supp := format [p] = "s" | p +:= 1);
              (char in string (format [p], loc int, s) | p +:= 1; true |
               p := q; false));
           proc copy = : ((¬ supp | put string (f, s [p1])); p1 +:= 1);
           proc intreal mould = : (? "r"; sign mould (fr [3]); int mould;
             (? "." | copy; int mould |: s [p1] = "." | p1 +:= 1);
             (? "e" | copy; sign mould (fr [6]); int mould));
           proc sign mould = (int sign) : (sign = 0 |
             (s [p1] = "−" |: ¬ value error of file | undefined) |
             s [p1] := (s [p1] = "+" | (sign | "+", ".") | "−");
             (? "z" | sign supp zero (s, p1, p1 + n) | n := 0);
             to n + 1 do copy; p +:= 1);
           proc int mould = : (l: (? "z" | bool zs := true; to n do
             (s [p1] = "0" ∧ zs | put string (file, "."); p1 +:= 1 |
               zs := false; copy); l); (? "d" | to n do copy; l));
           proc string mould = : while ? "a" do to n do copy;
           (fr [1] = 6 ∨ fr [1] = 9 | string mould |: intreal mould;
            fr [1] > 3 | p +:= 1; copy; intreal mould)
       end;

h) proc % edit choice = (file f, int c, string format, ref int p):
     (c > 0 | do insertion (f, format, p); p +:= 2; to c − 1 do
       (skip lit (format, p); format [p] = "," | p +:= 1 | undefined);
       do lit (f, format, p);
       while format [p] ≠ ")" do (p +:= 1; skip lit (format, p)); p +:= 1 |
       undefined);

i) proc % edit bool = (file f, bool b, string format, ref int p):
     (do insertion (f, format, p); (format [p + 1] = "(" | p +:= 2;
       (b | do lit (f, format, p); p +:= 1; skip lit (format, p) |
         skip lit (format, p); p +:= 1; do lit (f, format, p)) |
       put string (f, (b | "1" | "0"))); p +:= 1);
```

j)
```
proc % do insertion = (file f, string s, ref int p) :
   while (p > ⌈ s | false |: do align (f, s, p) | true | do lit (f, s, p)) do skip;
```

k)
```
proc % do align = (file f, string s, ref int p) bool :
   (int q = p; int n; replicator (s, p, n);
    (s [p] = "x" | to n do space (f); l |:
     s [p] = "y" | to n do backspace (f); l |:
     s [p] = "p" | to n do new page (f); l |:
     s [p] = "l" | to n do new line (f); l |:
     s [p] = "k" | char of bfile of f := n; l); p := q; false.
   l: p +:= 1; true);
```

l)
```
proc % do lit = (file f, string s, ref int p) bool :
   (int q = p; int n; replicator (s, p, n); (s [p] = """" |
     while (s [p +:= 1] = """" | s [p +:= 1] = """" | true) do
       put string (f, s [p]); true | p := q; false));
```

10.5.3.2. Formatted Input

a)
```
proc inf = (file file, tamrof tamrof, [ ] intype x) :
   (format (file, tamrof); in (file, x));
```

b)
```
proc in = (file file, [1 : ] intype x) :
   begin string format = format of file; ref int p = forp of file;
          for k to ⌈ x do
          ([1 : ] intype y = straightin x [k]; int q, j := 0;
           [1 : 14] int frame;
     rep: j +:= 1;
    step: while (exp insertion (file, format, p); p > ⌈ format | false |
          format [p] = ",") do p +:= 1; (j > ⌈ y | end);
          (p > ⌈ format | (¬ format end of file | p := 1); step);
          q := p; picture (format, q, frame);
          (frame [1] | int, real, real, compl, compl, string, intch, bool, char);
     int: (⟨(ref L int ii; (ii ::= y [j] |
           indit L int (file, ii, format, p, frame); rep))⟩); incomp;
    real: (⟨(ref L real xx; (xx ::= y [j] |
           indit L real (file, xx, format, p, frame); rep))⟩); incomp;
   compl: (⟨(ref L compl zz; (zz ::= y [j] |
           indit L compl (file, zz, format, p, frame); rep))⟩); incomp;
  string: (ref string ss; string t; (ss ::= y [j] |
           (frame [4] = 0 | get (file, ss) |
           indit string (file, t, format, p, frame); ss [@ 1] := t); rep));
    char: (ref char cc; string t; (cc ::= y [j] |
           indit string (file, t, format, p, frame);
           (ref char : cc) := t [1]; rep) | incomp);
   intch: (ref int ii; (ii ::= y [j] |
           indit choice (file, ii, format, p); rep)); incomp;
    bool: (ref bool bb; (bb ::= y [j] |
           indit bool (file, bb, format, p); rep));
```

```
      incomp: (value error of file | rep | undefined);
        end: skip)
        end;

c)  proc % indit L int = (file f, ref L int i, string format, ref int p,
      [ ] int fr) : (string t; indit string (f, t, format, p, fr);
        i := L string int (t, fr [2]));

d)  proc % indit L real = (file f, ref L real x, string format, ref int p,
      [ ] int fr) : (string t; indit string (f, t, format, p, fr);
        x := L string real (t));

e)  proc % indit L compl = (file f, ref L compl z, string format, ref int p,
        [ ] int fr) : (string t; int i; indit string (f, t, format, p, fr);
        z := (char in string ("⊥", i, t) |
          (L string real (t [1 : i − 1]) ⊥ L string real (t [i + 1 : ]))));

f)  proc % indit string = (file f, ref string t, string format, ref int p, [ ] int fr) :
      begin int n; bool supp; char k; string x; priority ! = 8;
              op ? = (char s) bool :
                (exp insertion (f, format, p); p > ⌈ format | false |
                  int q = p; replicator (format, p, n);
                  (supp := format [p] = "s" | p +:= 1);
                  (char in string (format [p], loc int, s) | p +:= 1; true |
                    p := q; false));
              op ! = (string s, char c) char :
                (char in string (next, loc int, s) | (supp | "" | k) |
                  char sugg := c; ((char error of f) (sugg) | sugg | undefined; c));
              op ! = (char s, c) char : (string : s) ! c;
              proc char next = char : (get string (f, k); k);
              proc intreal mould = : (? "r"; sign mould (fr [3]); int mould;
                (? "." | x +:= "." ! "."; int mould);
                (? "e" | x +:= "e10" ! "10"; sign mould (fr [6]); intmould));
              proc sign mould = (int sign) : (sign = 0 | x +:= "+" |
                int j := 0; (¬ ? "z" | n := 0);
                for i to n + 1 while next = "." do j := i;
                x +:= (k = "−" ∨ k = "+" ∧ sign = 1 | k |
                  (k ≠ "+" | j −:= 1; backspace (f)); "" ! "+");
                for i from j + 1 to n + 1 do x +:= "0123456789" ! "0");
              proc int mould = : (l: (? "z" | int j;
                for i to n while next = "." do j := i; backspace (f);
                from j to n do x +:= "0123456789" ! "0"; l);
                (? "d" | to n do x +:= "0123456789" ! "0"; l));
              proc string mould = :
                while ? "a" do to n do x +:= (supp | "." | next);
              (fr [1] = 6 ∨ fr [1] = 9 | string mould |: intreal mould;
               fr [1] > 3 | "⊥" ! "⊥"; intreal mould); t := x
        end;
```

```
g) proc % indit choice = (file f, ref int c, string format, ref int p):
     (exp insertion (f, format, p); p +:= 2; c := 1;
      while ask lit (f, format, p) do
        (c +:= 1; format [p] = "," | p +:= 1 | undefined);
      while format [p] ≠ ")" do (p +:= 1; skip lit (format, p));
      p +:= 1; exp insertion (f, format, p));
h) proc % indit bool = (file f, ref bool b, string format, ref int p):
     (exp insertion (f, format, p); (format [p + 1] = "(" | p +:= 2;
       (b := ask lit (f, format, p) | p +:= 1; skip lit (format, p) |:
        p +:= 1; ask lit (f, format, p) | undefined) | char k;
        get string (f, k); b := (k = "1" | true |: k = "0" | false));
      p +:= 1; exp insertion (f, format, p));
i) proc % exp insertion = (file f, string s, ref int p):
     while (p > ⌈ s | false |: do align (f, s, p) | true |
       exp lit (f, s, p)) do skip;
j) proc % exp lit = (file f, string s, ref int p) bool:
     (int q = p; int n; replicator (s, p, n); (s [p] = """" | int r = p;
       to n do (p := r; while (s [p +:= 1] = """" | s [p +:= 1] = """" |
       true) do (char k; get string (f, k); k ≠ s [p] | undefined)); true |
       p := q; false));
k) proc % ask lit = (file f, string s, ref int p) bool:
     (int c = char of bfile of f; int n; replicator (s, p, n);
       (s [p] = """" | int r = p; to n do (p := r;
          while (s [p +:= 1] = """" | s [p +:= 1] = """" | true) do
            (char k; get string (f, k); k ≠ s [p] | l)); true.
     l: while (s [p +:= 1] = """" | s [p +:= 1] = """" | true) do skip;
        char of bfile of f := c; false));
```

10.5.4. Binary Transput

a) **proc** % *to bin* = (**file** *f*, **simplout** *x*) [ ] **int**:
 c *a value of mode 'row of integral' whose lower bound is one, and whose upper bound depends on the value of 'f' and on the mode of the value of 'x'; furthermore, x = from bin (f, x, to bin (f, x))* **c**;

b) **proc** % *from bin* = (**file** *f*, **simplout** *v*, [ ] **int** *y*) **simplout**:
 c *a value, if one exists, of the mode of the actual-parameter corresponding to v, such that y = to bin (f, from bin (f, v, y))***c**;

{On some channels a more straightforward way of transput is available. Some properties of this binary transput depend on the particular implementation, others can be deduced from 10.5.4.}

10.5.4.1. Binary Output

```
a) proc write bin = ([ ] outtype x) : put bin (stand back, x);
b) proc put bin = (file file, [1 : ] outtype x):
      if bin possible [chan of file] ∧ opened of file ∧ put possible [chan of file]
   then   if ¬ set possible [chan of file] thef state def of file
```

```
        then (state get of file ∨ ¬ state bin of file | undefined)
        else state def of file := state bin of file := true;
             state get of file := false
          fi;
        for k to ⌈ x do ([1 : ] simplout y = straightout x [k];
          for j to ⌈ y do ([1 : ] int bin = to bin (file, y [j]);
            ref bfile b = bfile of file;
            ref int p = page of b, l = line of b, c = char of b;
            for i to ⌈ bin do (next plc (file); check plc (file);
              (book of b) [p, l, c] := bin [i]; c +:= 1;
              (p = lpage of b ∧ l = lline of b | (c > lchar of b | lchar of b := c) |
                  lpage of b := p; lline of b := l; lchar of b := c))))
    else undefined
      fi;
```

10.5.4.2. Binary Input

a) **proc** *read bin* = ([ ] **intype** *x*) : *get bin* (*stand back*, *x*);

b)
```
    proc get bin = (file file, [1 : ] intype x) :
        if bin possible [chan of file] ∧ opened of file ∧ get possible [chan of file]
    then    if ¬ set possible [chan of file] thef state def of file
        then (¬ state get of file ∨ ¬ state bin of file | undefined)
        else state def of file := state bin of file := state get of file := true
          fi; simplout y1;
        for k to ⌈ x do ([1 : ] intype y = straightin x [k];
          for j to ⌈ y do (y1 ::= y [j];
            [1 : 0 flex] int bin := to bin (file, y1); ref bfile b = bfile of file;
            for i to ⌈ bin do (next plc (file); check plc (file);
              bin [i] := (book of b) [page of b, line of b, char of b];
              char of b +:= 1);
            (<(ref L int ii; (ii ::= y [j] |
              (ref L int : ii) ::= from bin (file, ii, bin)))>);
            (<(ref L real xx; (xx ::= y [j] |
              (ref L real : xx) ::= from bin (file, xx, bin)))>);
            (<(ref L compl zz; (zz ::= y [j] |
              (ref L compl : zz) ::= from bin (file, zz, bin)))>);
            (ref string ss; (ss ::= y [j] |
              (ref string : ss) ::= from bin (file, ss, bin)));
            (ref char cc; (cc ::= y [j] |
              (ref char : cc) ::= from bin (file, cc, bin)));
            (ref bool bb; (bb ::= y [j] |
              (ref bool : bb) ::= from bin (file, bb, bin)))))
    else undefined
      fi;
```

{*But Eeyore wasn't listening. He was taking the balloon out, and putting it back again, as happy as could be. ...*
Winnie-the-Pooh, *A. A. Milne.*}

10.6. Standard Postlude

a) *lock (stand in); lock (stand out); lock (stand back)*

11. Examples

11.1. Complex Square Root

A declaration in which *compsqrt* is a procedure-with-[complex]-parameter-[complex]-mode-identifier (here [complex] stands for structured-with-real-field-letter-r-letter-e-and-real-field-letter-i-letter-m.):

a) **proc** *compsqrt* = (**compl** *z*) # *the square root whose real part is nonnegative of the complex number z* # **compl** :
b) **begin real** *x* = **re** *z*, *y* = **im** *z*;
c) **real** *rp* = *sqrt* ((**abs** *x* + *sqrt* (*x* ↑ 2 + *y* ↑ 2)) / 2);
d) **real** *ip* = (*rp* = 0 | 0 | *y* / (2 × *rp*));
e) (*x* ≥ 0 | *rp* ⊥ *ip* | **abs** *ip* ⊥ (*y* ≥ 0 | *rp* | −*rp*))
f) **end**

[complex]-calls {8.6.2} using *compsqrt*:

g) *compsqrt (w)*
h) *compsqrt (−3.14)*
i) *compsqrt (−1)*

11.2. Innerproduct 1

A declaration in which *innerproduct 1* is a procedure-with-integral-parameter-and-procedure-with-integral-parameter-real-parameter-and-procedure-with-integral-parameter-real-parameter-real-mode-identifier:

a) **proc** *innerproduct 1* = (**int** *n*, **proc** (**int**) **real** *x*, *y*) **real** :
comment *the innerproduct of two vectors, each with n components, x (i), y (i), i = 1, ..., n, where x and y are arbitrary mappings from integer to real number* **comment**
b) **begin long real** *s* := **long** *0*;
c) **for** *i* **to** *n* **do** *s* **plus leng** *x (i)* × **leng** *y (i)*;
d) **short** *s*
e) **end**

Real-calls {8.6.2} using *innerproduct 1*:

f) *innerproduct 1 (m, (***int** *j)* **real** : *x1 [j], (***int** *j)* **real** : *y1 [j])*
g) *innerproduct 1 (n, nsin, ncos)*

11.3. Innerproduct 2

A declaration in which *innerproduct 2* is a procedure-with-reference-to-row-of-real-parameter-and-reference-to-row-of-real-parameter-real-mode-identifier:

a) **proc** *innerproduct 2* = (**ref** [*1* :] **real** *a;* **ref** [*1* : ⌈ *a*] **real** *b)* **real** :
the innerproduct of two vectors a and b with equal number of elements
b) **begin long real** *s* := **long** *0;*
c) **for** *i* **to** ⌈ *a* **do** *s* +:= **leng** *a* [*i*] × **leng** *b* [*i*];
d) **short** *s*
e) **end**

Real-calls using *innerproduct 2*:

f) *innerproduct 2 (x1, y1)*
g) *innerproduct 2 (y2 [2], y2 [, 3])*

11.4. Innerproduct 3

A declaration in which *innerproduct 3* is a procedure-with-reference-to-integral-parameter-and-integral-parameter-and-procedure-real-parameter-and-procedure-real-parameter-real-mode-identifier:

a) **proc** *innerproduct 3* = (**ref int** *i,* **int** *n,* **proc real** *xi, yi)* **real** :
the innerproduct of two vectors whose n elements are the values of the expressions xi and yi and which depend, in general, on the value of i
b) **begin long real** *s* := **long** *0;*
c) **for** *k* **to** *n* **do** *(i := k; s +:=* **leng** *xi* × **leng** *yi);*
d) **short** *s*
e) **end**

A real-call using *innerproduct 3*:

f) *innerproduct 3 (j, 8, x1 [j], y1 [j + 1])*

11.5. Largest Element

A declaration in which *absmax* is a procedure-with-reference-to-row-of-row-of-real-parameter-and-reference-to-real-parameter-and-reference-to-integral-parameter-and-reference-to-integral-parameter-void-mode-identifier:

a) **proc** *absmax* = (**ref** [*1*:, *1*:] **real** *a,* # *result* # **ref real** *y,*
subscripts # **ref int** *i, k)* :
the absolute value of the element of greatest absolute value of the matrix a is assigned to y, and the subscripts of this element to i and k
c) **begin** *y* := −*1;*
d) **for** *p* **to** *1* **upb** *a* **do for** *q* **to** *2* **upb** *a* **do**
e) **if abs** *a* [*p, q*] > *y* **then** *y* := **abs** *a* [*(i := p), (k := q)*] **fi**
f) **end**

Void-calls using *absmax*:

```
g) absmax (x2, x, i, j)
h) absmax (x2, x, loc int, loc int)
```

11.6. Euler Summation

```
a) proc euler = (proc (int) real f, real eps, int tim) real:
       # the sum for i from 1 to infinity of f(i), computed by means of a suitably
         refined Euler transformation. The summation is terminated when the absolute
         values of the terms of the transformed series are found to be less than eps tim
         times in succession. This transformation is particularly efficient in the case
         of a slowly convergent or divergent alternating series #
b) begin int n := 1, t; real mn, ds := eps; [1 : 16] real m;
c)       real sum := (m [1] := f (1)) / 2;
d)       for i from 2 while (t := (abs ds < eps | t + 1 | 1)) ≤ tim do
e)         begin mn := f (i);
f)           for k to n do begin mn := ((ds := mn) + m [k]) / 2;
g)                                m [k] := ds end;
h)           sum plus (ds := (abs mn < abs m [n] ∧ n < 16 |
i)                            n plus 1; m [n] := mn; mn / 2 | mn))
j)         end;
k)       sum
l) end
```

A call using *euler*:

```
m) euler ((int i) real: (odd i | -1 / i | 1 / i), 1₁₀-5, 2)
```

11.7. The Norm of a Vector

```
a) proc norm = (ref [1 : ] real a) real:
       # the euclidean norm of the vector a #
b) (long real s := long 0;
c) for k to upb a do s plus leng a [k] ↑ 2;
d) short long sqrt (s))
```

For a use of *norm* as a call, see 11.8.e.

11.8. Determinant of a Matrix

```
a) proc det = (ref [1 :, 1 : ] real a; ref [1 : upb a] int p) real:
b)   if upb a = 2 upb a
c)   then int n = upb a;
       # the determinant of the square matrix of a order n by the method of Crout
         with row interchanges: a is replaced by its triangular decomposition l × u
         with all u [k, k] = 1. The vector p gives as output the pivotal row indices;
         the k-th pivot is chosen in the k-th column of l such that abs l [i, k] / row
         norm is maximal. #
```

```
d)      [1 : n] real v; real d := 1, s, pivot;
e)      for i to n do v [i] := norm (a [i]);
f)      for k to n do
g)        begin int k1 = k - 1; ref int pk = p [k]; real r : = - 1;
h)        ref [,] real al = a [, 1 : k1], au = a [1 : k1];
i)        ref [ ] real ak = a [k], ka = a [, k],
j)          alk = al [k], kau = au [, k];
k)        for i from k to n do
l)          begin ref real aik = ka [i];
m)          if (s := abs (aik -:= innerproduct 2 (al [i], kau)) / v [i]) > r
n)          then r := s; pk := i fi
o)          end;
p)        v [pk] := v [k]; pivot := ka [pk]; ref [ ] real apk = a [pk];
q)        for j to n do
r)          begin ref real akj = ak [j], apkj = apk [j];
s)          r := akj; akj := if j ≤ k then apkj
t)            else (apkj - innerproduct 2 (alk, au [, j])) / pivot fi;
u)          if pk ≠ k then apkj := -r fi
v)          end;
w)        d ×:= pivot
x)        end;
y)      d
z)    fi
```

A call using *det*:

```
aa) det (y2, i1)
```

11.9. Greatest Common Divisor

```
a)  proc gcd = (int a, b) int :
      # the greatest common divisor of two integers #
b)    (b = 0 | abs a | gcd (b, a ÷: b))
```

A call using *gcd*:

```
c)  gcd (n, 124)
```

11.10. Continued Fraction

```
a)  op / = ([1: ] real a; [1: upb a] real b) real :
      comment the value of a / b is that of the continued fraction
      a1 / (b1 + a2 / (b2 + ... an / bn) ... ) comment
b)    (upb a = 0 | 0 | a [1] / (b [1] + a [2 : ] / b [2 : ]))
```

A formula using /:

```
c)  x1 / y1
```

{The use of recursion may often be elegant rather than efficient as in the recursive procedure 11.9 and the recursive operation 11.10. See, however, 11.11 and 11.14 for examples in which recursion is of the essence.}

11.11. Formula Manipulation

```
a)  begin union form = (ref const, ref var, ref triple, ref call);
b)  struct const = (real value);
c)  struct var = (string name, real value);
d)  struct triple = (form left operand, int operator, form right operand);
e)  struct function = (ref var bound var, form body);
f)  struct call = (ref function function name, form parameter);
g)  int plus = 1, minus = 2, times = 3, by = 4, to = 5;
h)  heap const zero, one; value of zero := 0; value of one := 1;
i)  op = = (form a, ref const b)
      bool : (ref const ec; (ec ::= a | ec :=: b | false));
j)  op + = (form a, b)
      form : (a = zero | b |: b = zero | a | triple := (a, plus, b));
k)  op − = (form a, b) form : (b = zero | a | triple := (a, minus, b));
l)  op × = (form a, b) form : (a = zero ∨ b = zero | zero |: a = one | b
                               |: b = one | a | triple := (a, times, b));
m)  op /  = (form a, b) form : (a = zero ∧ ¬ (b = zero) | zero
                               |: b = one | a | triple := (a, by, b));
n)  op ↑ = (form a, ref const b) form : (a = one ∨ (b :=: zero) | one
                               |: b :=: one | a | triple := (a, to, b));
o)  proc derivative of = (form e, # with respect to # ref var x) form :
p)  begin ref const ec; ref var ev; ref triple et; ref call ef;
q)    case ev, et, ef ::= e in
r)      # ev # (ev :=: x | one | zero),
s)      # et # begin form u = left operand of et, v = right operand of et;
t)               form udash = derivative of (u, # with respect to # x),
u)                    vdash = derivative of (v, # with respect to # x);
v)               case operator of et in
w)                 udash + vdash, udash − vdash,
x)                 u × vdash + udash × v, (udash − et × vdash) / v,
y)                 (ec ::= v | v × u ↑
                   (heap const c; value of c := value of ec − 1; c) × udash)
               esac
             end ,
z)      # ef # begin ref function f = function name of ef;
aa)              form g = parameter of ef;
ab)              ref var y = bound var of f;
ac)              heap function fdash := (y, derivative of (body of f, y));
```

```
ad)            (call := (fdash, g)) × derivative of (g, x)
            end
ae)     out # ec # zero
        esac # ev, et, ef, ec #
      end # derivative #;
af) proc value of = (form e) real:
ag)     begin ref const ec; ref var ev; ref triple et; ref call ef;
ah)       case ec, ev, et, ef ::= e in
ai)         # ec # value of ec,
aj)         # ev # value of ev,
ak)         # et # begin real u = value of (left operand of et),
al)                           v = value of (right operand of et);
am)                  case operator of et in
an)                    u + v, u − v, u × v, u / v, exp (v × ln (u)) esac
                  end,
ao)         # ef # begin ref function f = function name of ef;
ap)                  value of bound var of f := value of (parameter of ef);
aq)                  value of (body of f)
                  end
          esac # ec, ev, et, ef #
        end # value of #;
ar) heap form f, g; heap var a := ("a", ∿), b := ("b", ∿), x := ("x", ∿);
as) start here: read ((value of a, value of b, value of x));
at) f := a + x / (b + x); g := (f + one) / (f − one);
au) print ((value of a, value of b, value of x,
            value of (derivative of (g, # with respect to # x))))
    end # example of formula manipulation #
```

11.12. Information Retrieval

```
a)  begin  mode ra = ref auth, rb = ref book,
           struct auth = (string name, ra next, rb book),
                  book = (string title, rb next);
b)         ra auth, first auth := nil, last auth; rb book;
c)         string name, title; int i; file input, output;
d)         open (input,, remote in); open (output,, remote out);
e)         outf (output, $p
f)                "to.enter.a.new.auther,.type.""author"",.a.space,.and.his.
                  name."l
g)                "to.enter.a.new.book,.type.""book"",.a.space,.the.name.of.the.
                  author,.a.new.line,.and.the.title."l
h)                "for.a.listing.of.the.books.by.an.author,.type.""list"",.a.space,.
                  and.his.name."l
```

```
i)                  "to.find.the.author.of.a.book,.type.""find"",.a.new.line,.and.the.
                    title."l
j)                  "to.end,.type.""end""""al$, ".");
k)          proc update =
l)            : if (ra : first auth) :=: nil
m)            then auth := first auth := last auth := auth := (name, o, o)
n)            else auth := first auth; while (ra : auth) :≠: nil do
o)                    (name = name of auth | known | auth := next of auth);
p)                  last auth := next of last auth := auth := auth :=
q)                    (name, o, o); known: skip
                fi # end declaration prelude sequence #;
r)    client: inf (input, $c("author", "book","list","find","end",""), x30al,
                       80al$, i);
s)          case i in author, publ, list, find, end, error esac;
t)   author: in (input, name); update; client;
u)     publ: in (input, (name, title)); update;
v)            if (rb : book of auth) :=: nil
w)          then book of auth := book := (title, o)
x)          else book := book of auth; while (rb : next of book) :≠: nil do
y)                  (title = title of book | client | book := next of book);
z)                (title ≠ title of book | next of book := book := (title, o))
aa)           fi; client;
ab)     list: in (input, name); update;
ac)         outf (output, $p"author:."30all$, name);
ab)           if (rb : book of auth) :=: nil
ae)         then put (output, "no.publications")
af)         else while (rb : book) :≠: nil do
ag)                 begin if line number (output) = max line [remote out]
ah)                     then outf (output, $41k"continued.on.next.page"p
                                  "author:."30a41k"continued"ll$, name)
ai)                       fi; outf (output, $80al$, title of book);
aj)                     book := next of book
                    end
ak)           fi; client;
al)    find: in (input, (loc string, title)); auth := first auth;
am)         while (ra : auth) :≠: nil do
an)           begin book := book of auth;
ao)                 while (rb : book) :≠: nil do
ap)                       if title = title of book
aq)                   then outf (output, $l"author:."30a$, name of auth); client
ar)                    else book := next of book
as)                       fi; auth := next of auth
at)              end;
```

```
au)          outf (output, $l"unknown"l$,); client;
av)     end: put (output, (new page, "signed.off", close));
aw)          close (input).
ax)   error: put (output, (new line, "mistake,.try.again."));
ay)          new line (input); client
    end
```

11.13. Cooperating Sequential Processes

```
a) begin int nmb magazine slots, nmb producers, nmb consumers;
b)    read ((nmb magazine slots, nmb producers, nmb consumers));
c)    [1: nmb producers] file infile, [1: nmb consumers] file outfile;
d)    for i to nmb producers do open (infile [i],, inchannel [i]);
        # inchannel and outchannel are defined in a surrounding range #
e)    for i to nmb consumers do open (outfile [i],, outchannel [i]);
f)    mode page = [1: 60, 1: 132] char;
g)    [1: nmb magazine slots] ref page magazine;
h)    int # pointers of a cyclic magazine # index := 1, exdex := 1;
i)    sema full slots = | 0, free slots = | nmb magazine slots,
j)      in buffer busy = | 1, out buffer busy = | 1;
k)    proc par call = (proc (int) p, int n) # calls n incarnations of p in
l)      parallel #: (n>0 | par (p (n), par call (p, n - 1)));
m)    proc producer = (int i) : do (heap page page; get (infile[i], page);
n)      ↓ free slots; ↓ in buffer busy;
o)      magazine [index] := page; index ÷::= nmb magazine slots +:= 1;
p)      ↑ full slots; ↑ in buffer busy);
q)    proc consumer = (int i) : do (page page;
r)      ↓ full slots; ↓ out buffer busy;
s)      page := magazine [exdex]; exdex ÷::= nmb magazine slots +:= 1;
t)      ↑ free slots; ↑ out buffer busy; put (outfile [i], page));
u)    par (par call (producer, nmb producers),
           par call (consumer, nmb consumers))
   end
```

11.14. Towers of Hanoi

```
a) begin proc p = (int me, de, ma) :
b)         (ma>0 | p (me, 6 - me - de, ma - 1);
c)                 out (stand out, (me, de, ma));
                   # move from peg 'me' to peg 'de' piece 'ma' #
d)                 p (6 - me - de, de, ma - 1));
e)         for k to 8 do (outf (stand out, $l"k.=."dl,
                      n((2↑k+15)÷16)(2(2(4(3(d)x)x)x)l)$, k);
f)                    p (1, 2, k))
     end
```

12. Glossary

12.1. Technical Terms

Given below are the locations of the first, and sometimes other, instructive appearances of a number of words which, in Chapters 1 up to 10 of this Report, have a specific technical meaning. A word appearing in different grammatical forms (e.g., "assign", "assigned", "assignment") is given once, usually as infinitive (e.g., "assign").

{Denn eben, wo Begriffe fehlen,
Da stellt ein Wort zur rechten Zeit sich ein.
Faust, J. W. von Goethe.}

12.2. Paranotions

Given below are the indicators of the rules yielding production rules for the originals of the given paranotions and other protonotions or giving instructive appearances of, or representations for, the given symbols. Ordinary type font without hyphens is used in order to shorten the text by using hyphens in a conventional way.

GPSR Compliance

The European Union's (EU) General Product Safety Regulation (GPSR) is a set of rules that requires consumer products to be safe and our obligations to ensure this.

If you have any concerns about our products, you can contact us on ProductSafety@springernature.com

In case Publisher is established outside the EU, the EU authorized representative is:

Springer Nature Customer Service Center GmbH
Europaplatz 3
69115 Heidelberg, Germany

Batch number: 10704763

Printed by Printforce, the Netherlands